AF517261

A Field Guide to Flowering Plants of The Mekal Hills, Central India

Authors

Mr. Shivaji Chaudhry is currently working as an Assistant Professor at the Department of Environmental Science, Indira Gandhi National Tribal University. He has started his research life from G.B. Pant Institute of Himalayan Environment and Development as a research scholar and Technician. He has been recipient of number of National and International Fellowship. He has contributed 8 book chapters, 15 research papers, 54 IUCN redlist assessments. He has also completed 5 national and international consultancies. He contributed 2 new species of freshwater fishes and also reviewed 84 freshwater fishes of Eastern Himalaya. He has interests in biodiversity, taxonomy and ecology of plants and animals.

Dr. Naveen Kumar Sharma is presently serving as a Professor at the Department of Botany. He did his M.Sc. in Botany with first class from the Banaras Hindu University and obtained his Doctoral degree from Jiwaji University, Gwalior. He has been in teaching profession since 2001. He did his post-doctoral training from the University of Austin, Texas USA. He has authored 42 research papers and 3 books, and has completed 2 projects. He is a life member of International Organization of Plant Proteomics, Japan.

Dr. Nidhan Singh obtained his Masters and Doctorate in Botany from Kurukshetra University, Kurukshetra, Haryana (India). He started his career as Assistant Professor at the Department of Botany, I.B. (PG) College, Panipat in 2006. He is presently serving as a senior Assistant Professor and Head, Department of Botany and Biotechnology. He is actively contributing to the floristic research of India through his regular explorations of Western Himalayas and other parts of country. He is one of the most active members and moderators of the efloraofindia (Google group) and his contributions can be seen by visiting https://sites.google.com/site/efloraofindia. He has also contributed for many pages to http://www.flowersofindia.net. Dr. Nidhan Singh has published many research papers in journals of repute and presented his research papers in many National and International conferences.

A Field Guide to Flowering Plants of The Mekal Hills, Central India

Shivaji Chaudhry
Naveen Kumar Sharma
Nidhan Singh

Published by

SCIENTIFIC PUBLISHERS (INDIA)
5 A, New Pali Road
P.O. Box 91
Jodhpur - 342 001 INDIA
E-mail: info@scientificpub.com
Website: http://www.scientificpub.com

ISBN: 978-93-88812-04-7 (Hardbound)
978-93-88812-06-1 (E-ISBN)

Printed in India

इंदिरा गाँधी राष्ट्रीय जनजातीय विश्वविद्यालय
Indira Gandhi National Tribal University
Inspiring Students, Empowering Society

www.igntu.ac.in

(संसद के अधिनियम द्वारा स्थापित केन्द्रीय विश्वविद्यालय)
(A Central University Established by an Act of Parliament)

Prof. T.V. Kattimani
Vice-Chancellor

Ref. No. / IGNTU/
Date: 15.03.2019

Foreword

Ever since I took over as the Vice-Chancellor of the Indira Gandhi National Tribal University (IGNTU), Amarkantak (Madhya Pradesh), the country's first national tribal university, it has been my mission to encourage my young colleagues to take up the issues that are not only topical but in either ways related to the vast but little documented traditional knowledge of the 8.2% tribal population of the country. The University is located amidst the picturesque setting where Vindhyan and Satpura ranges meet, with the Mekal Hills being the fulcrum. This is the birth place of mighty rivers Narmada and Son. The forest-claded area is dominated by the Sal trees. One of the objectives entrusted upon the IGNTU is to explore, document and promote proper management and justful use of the natural resources, including biological resources, of the area heavily inhabited by the native tribal population.

In addition to the above-stated academic underpinnings, throughout the year, the area receives visitors from various parts of the country, primarily due to the religious importance of Amarkantak in Hindu mythology. In this context, it is my pleasure to welcome **"Field Guide to Flowering Plants of The Mekal Hills"** written by my colleagues Mr. Shivaji Chaudhry (Assistant Professor at the Department of Environmental Science, IGNTU, Amarkantak) and Prof. Naveen Kumar Sharma (Department of Botany, IGNTU, Amarkantak) along with Dr. Nidhan Singh from IB (PG) College, Paniput, Haryana. I hope that the book will be greatly received by naturalists, foresters, academicians and nature lovers. For this, the authors deserve hearty congratulation.

(Prof. T.V. Kattimani)

अमरकंटक, जिला–अनूपपुर, म.प्र. – 484887
Phone No. / Fax No. - 07629 - 269710 (O)

Amarkantak, Dist. Anuppur, M.P. - 484887
e-mail vc@igntu.ac.in

PREFACE

This is a small endeavor from our side to document the plant wealth of the Mekal Hills. This densely forested and thinly populated range is birth place to several streams and rivers including Narmada and Son Rivers. The hills are inhabited by many tribal populations notably amongst them are - the Baigas and the Gonds. The book is an outcome of a minor project sanctioned by the Madhya Pradesh State Biodiversity Board, to one of us, Mr. Shivaji Chaudhary. It aims to familiarize the flora of the hills amongst botanists, tourists, forest people and nature lovers without bothering much of taxonomic nitty-gritty. The main strength of the book lies in pictorial depiction of the taxa and their description in simplest words.

Shivaji Chaudhry
Naveen Kumar Shrama
Nidhan Singh

IGNTU, Amarkantak & Panipat, Haryana
Dated 15 March 2019

CONTENTS

List of Photos

List of Photos

List of Photos

List of Photos

List of Photos

INTRODUCTION AND STUDY AREA

The flowering plants are among the most evolved form of plants, as is evidenced from the numerous advantages they have, like sexual reproduction, seed production, double fertilization and a reduced female gametophyte. Being the most advanced plant domain, they have evolved about 130 million year ago during the Jurassic and Cretaceous era. Based on the seed character, they are broadly divided into monocots (with single cotyledon) and dicots (with two cotyledons). According to famous evolutionary theorist Armen Thakhtajan (1969), the cradle of flowering plant lies between India and Fiji. India is one of the topographically and climatically most diverse countries, with an area of 329 million hectares and the forest cover of 21% (FSI 2011). It is also home to about 17,296 species of angiosperms (Singh and Dash, 2014). The central India comprises of states of Madhya Pradesh, Chhattisgarh and parts of Maharashtra. Amongst these, the state of Madhya Pradesh has forest cover of 28.27%, while Chhattisgarh has 46.04% land covered under forests. Most of the forests in the region are of mixed deciduous type and are lush green during the monsoon time from June to September. The temperature of the area ranges from 11°C-45°C while the precipitation ranges from 800-1800 mm per annum. A total of 2000 species of angiosperms have been recorded from the region but most of these surveys have not been conducted on regular basis (Roy *et al.,* 1992). Topography seems to play an important role in floral dispersal and speciation as undulating mountains of the eastern and southern parts of Central India harbor maximum biodiversity. The Vindhya, Satpura and Mekal hills criss-cross the entire region, and these mountains are a treasure of natural floristic diversity. The significance of area is further enhanced by the fact that the region is not only rich in flora and fauna, but also equally enriched by the diverse tribal communities inhabiting the area.

Mekal hill is one such unique mountain, which covers the twin states of Chhattisgarh and Madhya Pradesh. It is one of the richest biodiversity treasure troves as it covers protected areas like Bhoramdeo Wildlife Sanctuary, Achanakmar-Amarkantak Biosphere reserve and Kanha National Park. The elevation of this hills ranges from 100-1100 m which gives rise to unique floral elements in the region. The landscape of Satpura-Mekal extends for a distance of about five hundred kilometers. The whole extent of this hill range touches the districts of Anuppur, Balaghat, Bilaspur, Mandla, Dindori and Kabirdham of

Madhya Pradesh and Chattisgarh. The Mekal hill region runs more than 500 km and is having Achanakmar-Amarkantak Biosphere reserve on one side and the other side runs up to Melghat Tiger Reserve in Maharashtra. Most of the botanical survey in the Central Indian region was carried out under Botanical Survey of India, Allahabad. One of the most famous floristically rich provinces is Kanha National Park which covers the districts of Mandla and Balaghat. About 200 species of Angiosperms including more than 70 species of trees have been recorded in this National Park. The forest is mixed deciduous type and it comprises of Sal (*Shorea robusta*) as dominant species, the other major species of trees are *Terminalia alata* (Saja), *Lagerstroemia parviflora* (Lendia), *Anogeissus latifolia* (Dhawa), *Diospyros melanoxylon* (Tendu), *Butea monosperma* (Palash) etc. Not much of information is available on floral aspect of Bhoramdeo Wildlife Sanctuary. However, this undulating topography is having mixed deciduous forest having trees of Sal, Tinsa, Saja, Haldu (*Adina cordifolia*), Lendia etc. Among these localities, perhaps one of the best studied areas is Achanakmar-Amarkantak Biosphere Reserve. It consists of about 1500 plant species distributed in 151 plant families. Some of the important species of angiosperms are *Thalictrum foliolosum*, *Dillenia pentagyna*, *Cocculus hirsutus*, *Flacourtia indica*, *Talinum portulacifolium*, *Tamarix aphylla*, *Abelmoschatus moschata*, *Hibiscus sabdariffa*, *Corchorus fascicularis*, *Biophytum sensitivum* etc.

Keeping in the mind the immense biodiversity potential of the Mekal region, a project proposal to document the angiosperms of this region was formulated. The main aim was to document the flora in form of literature, also supplemented with field photography. This would act as a field guide for the educationist, researchers, foresters and students as a whole. The objectives of the project were:

(i) To document and keep photographic evidence of the plants of this region.

(ii) To find out their usefulness in daily life of the local tribals of this region.

(iii) To publish and disseminate the knowledge to various stakeholders e.g. educational institutions, research institutes, forest department in the form of book and research articles.

The realization of these objectives have taken the form of this book, providing an insight into major flowering plants of the region.

MATERIALS AND METHOD

As a main aspect of the project was to document the angiosperm diversity of the region, many surveys were conducted especially during the major flowering and fruiting seasons. The surveys were carried out during May 2016 to December 2017. Visits to the protected areas were carried out and the flowering or fruiting plants were photographed with the help of a digital camera (Sony H400). For ensuring maximum collection of floral elements, mostly protected areas i.e.

Achanakmar-Amarkantak Biosphere Reserve, Bhoramdeo Wildlife Sanctuary and Kanha National Parks were covered. Surveys were also conducted outside these protected areas in the districts of Anuppur, Balaghat, Bilaspur, Mandla and Kabirdham of Madhya Pradesh and Chattisgarh. Monthly visits were carried out across all four seasons to cover the maximum flowering season. Plants were identified using the field guides such as: Flowers of the Himalaya (Polunin and Stainton, 1984), A Tree Spotters : Jungle tree field guides of Central India (Krishen, 2014), Common Indian wild Flowers (Khemikar, 2000), Wild edible fruit plants of Eastern India (Mahapatra and Panda, 2009) and taxonomic literature such as Flora of Madhya Pradesh (Roy *et al.*, 1992). Care was taken to collect the plants both in digital format by photography, as well as, in form of herbarium specimens, which are deposited at the department of environmental science, Indira Gandhi National Tribal University, Amarkantak. The identified plants were further checked for latest taxonomic update on ITIS catalogue of life (www.catalogueoflife.org), GRIN, TPL, efloraofindia. These plants were then cross checked with IUCN red list for threat assessment (www.iucnredlist.org).

RESULTS

The study resulted in recording of 410 species of flowering plants (angiosperms), distributed in 321 genera and 92 families (Figure 1). Among this, 70% of the plants were found near Achanakmar-Amarkantak Biosphere Reserve, 25% were found near Kanha National Park and remaining 5% were found near Bhoramdeo Wildlife Sanctuary.

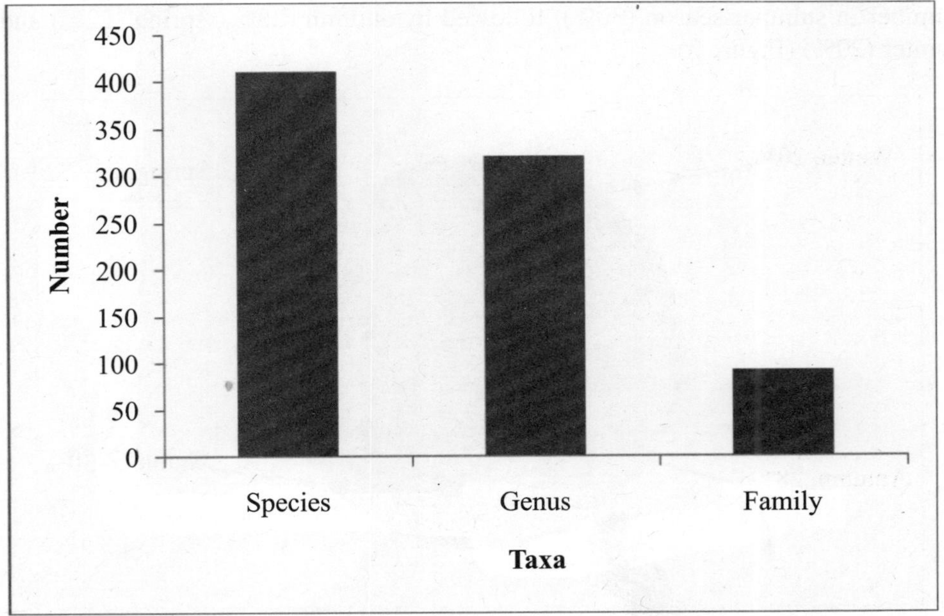

Figure 1: Proportion of Various Angiospermic Taxa in the Study Area

All the plants were subjected to habit-wise classification (herbs, shrubs, creepers, epiphyte etc.). Accordingly, about 222 species of herbs, 100 tree species, 51 shrubs, 35 creepers and 2 epiphytes were documented during the survey. The dominant habit of the plants were herbs 54% and trees 24%, respectively (Figure 2)

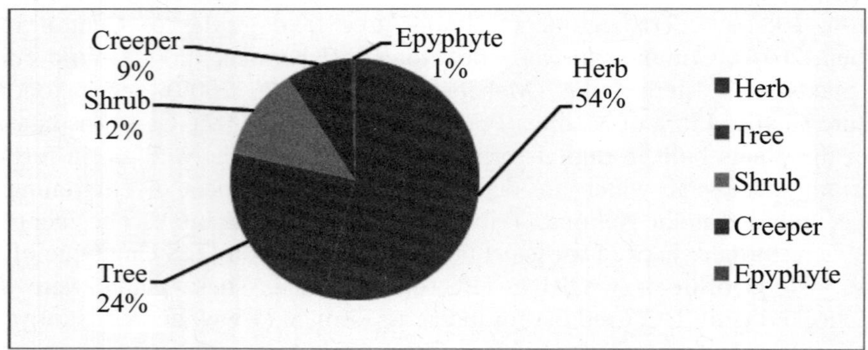

Figure 2: Habit-wise Distribution of Flowering Plants of Mekal Hills

The classification of the flowering plants recorded from the area was further extended to the seasonal groups, to know the important phenological activities like flowering and fruiting in different seasons. The four seasons identified were Spring (March-May) Summer (June-August), Autumn (September-November) and Winter (December-February). Overlapping phenologies were recorded in many species, but percentage distribution of phenophase revealed maximum number in summer season (30%), followed by autumn (28%), spring (22%) and winter (20%) (Figure 3).

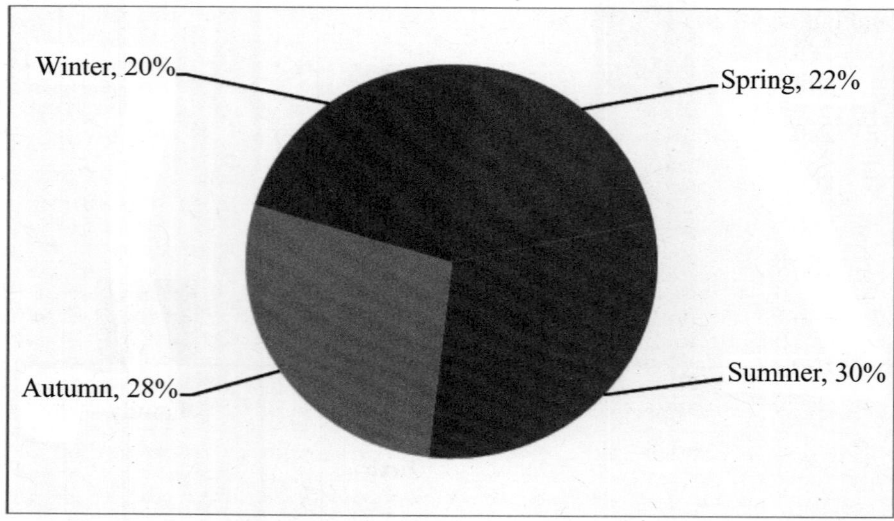

Figure 3: Seasonal reproductive phenophase of Angiosperms from Mekal Hills.

The threat and vulnerability assessment of plant species was carried out with reference to global IUCN red listing as Not Assessed (NA), Least Concerned (LC), Data Deficient (DD) and Vulnerable (Figure 4). The data showed that the largest fraction of the plants recorded from area i.e. 352 (85.57%) were of Not Assessed 53 (12.93%). Least concerned 4 (0.98%). Data deficient and 1 (0.24%) of Vulnerable category.

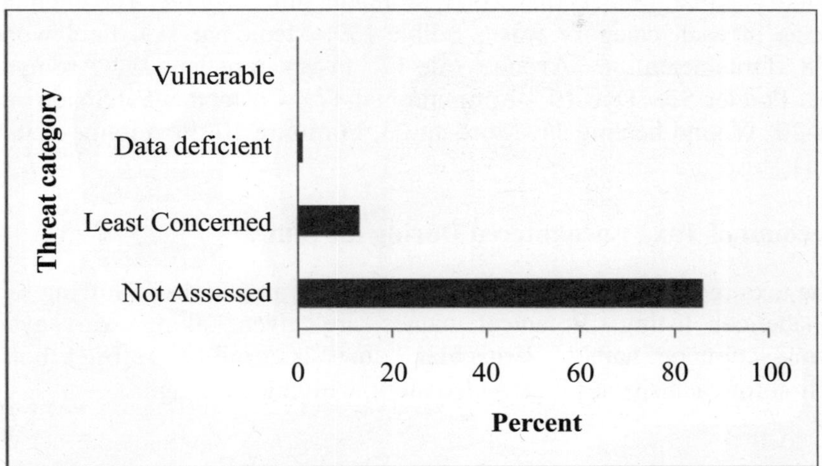

Figure 4: Proportionate Distribution of Angiosperms in IUCN threat assessment list

Threat listed plants were further classified using IUCN redlist with respect to their population status in wild. The population criteria were divided into stable, increasing, decreasing and unkown (Figure 5). The results with respect to their percentage category were : stable 27 (52.9%), increasing 3 (5.9%), decreasing 1 (2%) and unknown 20 (39.2%).

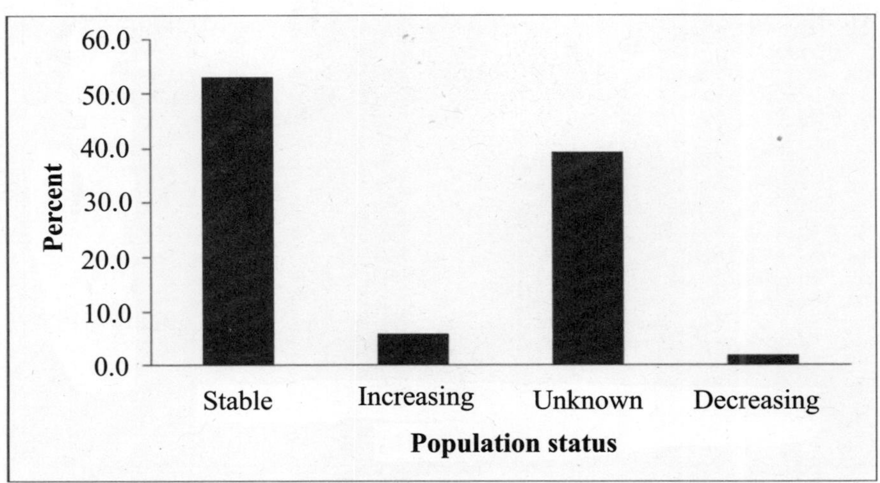

Figure 5: Population status of flowering plants.

Factors associated with threats to the plants were also listed as per IUCN redlist, a total of 9 threat categories were identified, out of them, threats due to unknown reasons was found for 45 species which amounted to 83%, followed by timber extraction (2 species, 3.7%). For remaining seven threat categories, only a single species (1.85%) was found to contribute.

All the recorded species were also subjected to classification based on utility, like - edible, timber, fuel wood, aromatic, gum, dye etc. The enumeration of species in each category was : Edible-139, Medicinal-319, Fuel wood-17, Fiber-28, Ornamental-86, Avenue tree-17, Insect repellent-17, Aromatic-25, Gum-5, Fodder-53, Dye-19, Antimicrobial-17, Cosmetics-5, Shade tree-3, Timber-20, Wound healing-11, Mordant-03, Furniture-10, Phytoreme-diation-04 (Table 1).

The Account of Taxa Encountered During the Study

The taxa recorded during the study are enumerated in the following section, in alphabetical listing. Botanical names are given, along with available synonyms, common names, vernacular names and utility. A brief botanical description for each species is provided along with digital image.

■■■■

TAXONOMY AND DESCRIPTION

1. Scientific Name: *Abelmoschus moschatus* (L.) Medic.

Synonym: *Hibiscus moschatus* (L.) Salisb.

Family: Malvaceae

Common Name: Musk Mellow

Vernacular Name: Ban Bhedi

Habitat: Humid tropical areas

Uses: Edible, Perfumery, Adsorbent, Medicinal

IUCN Threat listing: Not Assessed (NA)

Description: A compact, tender, perennial herb; Leaves palmately 3-7-lobed, lobes narrowly linear, coarsely serrate, sparsely hirsute, apex acuminate, base cordate; Flowers solitary, axillary, pedicel hirsute, epicalyx lobes 6-10, narrow oblong; Capsules 6-8 cm long, oblong-lanceolate, strongly 5-angular, appressed hairy, blackish when ripe.

Flowering and Fruiting: July-September

2. Scientific Name: *Abroma angusta* (L.) L. *f.*

Synonym: *Theobroma angusta* L.

Family: Malvaceae

Common Name: Cocoa tree

Vernacular Name: Ulatkambal

Habitat: Open tropical areas

Uses: Medicinal

IUCN Threat listing: Not Assessed (NA)

Description: A large, spreading shrub, or a small tree, with fibrous bark and irritant hairs; Leaves ovate-oblong, long-pointed, with a heart-shaped base, 10-21 cm long; Flowers maroon, up to 5 cm across, looking down, in few-flowered clusters in leaf axils; Capsule papery, 5-winged, cut-off at the tip.

Flowering and Fruiting: February-June

3. Scientific Name: *Acacia auriculiformis* A. Cunn. *ex* Benth.

Synonym: *Acacia moniliformis* Griseb.

Family: Leguminosae

Common Name: Earleaf Acacia

Vernacular Name: Australian Babul

Habitat: Open humid tropical areas

Uses: Avenue tree, Fuelwood

IUCN Threat listing: Least Concerned (LC)

Description: A medium to tall tree, heavily branched, with a short bole, bark grey or brown, sometimes blackened at the base; Phyllodes falcate, glabrous, greyish-green and thinly textured; Flowers light-golden, 5-merous, bisexual, tiny, sessile, fragrant, calyx tubular; Pods strongly curved to form an open coil, flat, flexible but woody, glaucous, transversely veined with undulate margins.

Flowering and Fruiting: March-December

4. Scientific Name: *Acalypha hispida* Burm. *f.*

Synonym: *Ricinocarpus hispidus* (Burm. *f.*) Kuntze

Family: Euphorbiaceae

Common Name: Chenille plant

Vernacular Name: Kuppi

Habitat: Open humid tropical areas

Uses: Ornamental, Hedge, Medicinal

IUCN Threat listing: Not Assessed (NA)

Description: A sparsely branched shrub, can grow 6-12 ft; Leaves evergreen, oval, pointed on the tips; Pistillate flowers purple, bright red or crimson, clustered in velvety catkins.

Flowering and Fruiting: July-October

5. Scientific Name: *Acalypha wilkesiana* Müll. Arg.

Synonym: *Ricinocarpus wilkesianus* (Müll.Arg.) Kuntze

Family: Euphorbiaceae

Common Name: Copper Leaf

Vernacular Name: Kuppi Kholi

Habitat: Open humid tropical areas

Uses: Ornamental, Hedge, Medicinal

IUCN Threat listing: Not Assessed (NA)

Description: An erect or spreading, evergreen, often suckering shrub, 2-4 m tall; Leaves variegated or brown-coloured, often twisted and aberrant; Flowers small and inconspicuous, in 12-25 cm long, somewhat drooping, green racemes often hidden in the foliage.

Flowering and Fruiting: November-July

6. Scientific Name: *Achyranthes aspera* L.

Synonym: *Achyranthes canescens* R. Br.

Family: Amaranthaceae

Common Name: Prickly chaff flower

Vernacular Name: Chirchita

Habitat: Open tropical grasslands

Uses: Medicinal

IUCN Threat listing: Not Assessed (NA)

Description: Perennial, erect herb; Stem branched, up to 1.2 m high; Leaves opposite, elliptic, obovate or sub-orbicular, pubescent, apex acute-acuminate, base cuneate-rounded; Flowers greenish-white, deflexed, in axillary or terminal spikes; Urticles oblong cylindric.

Flowering and Fruiting: August-December

7. Scientific Name: *Acmella oleracea* (L.) R.K. Jansen

Synonym: *Anacyclus pyrethraria* (L.) Spreng

Family: Asteraceae

Common Name: Para cress

Vernacular Name: Akarkara

Habitat: Open humid tropical areas

Uses: Edible, Medicinal

IUCN Threat listing: Not Assessed (NA)

Description: Erect or ascending anuual herb; Stem more or less pubescent, 30-60 cm high; Leaves opposite, stalked, ovate-lanceolate, acute, entire or toothed; Flowers white or rusty yellow, heads conical, solitary on long stalks, ovoid; Achenes oblong, glabrous or scarbid.

Flowering and Fruiting: Throughout the Year

8. Scientific Name: *Adenanthera microsperma* Teijsm. & Binn.

Synonym: *Adenanthera pavonina* L.

Family: Leguminosae

Common Name: Coral wood

Vernacular Name: Ratan Gunj

Habitat: Open humid tropical areas

Uses: Jewellery, Edible, Shade tree

IUCN Threat listing: Not Assessed (NA)

Description: Medium to large-size deciduous tree, erect, having dark brown to greyish bark; Leaves bipinnate, leaflets alternate, oval-oblong, base asymmetric, apex blunt; Flowers borne in narrow spike-like racemes, at branch ends, small, creamy-yellow, fragrant; Seed pods split open to reveal the bright red seeds, thus the common Name bead tree.

Flowering and Fruiting: March-August

9. Scientific Name: *Adina cordifolia* (Roxb.) Brandis

Synonym: *Haldina cordifolia* (Roxb.) Ridsdale

Family: Rubiaceae

Common Name: Yellow teak

Vernacular Name: Hardu

Habitat: Open tropical areas

Uses: Fuel wood, Furniture, Medicinal

IUCN Threat listing: Not Assessed (NA)

Habit: Tree

Description: Large, deciduous tree Stem buttressed with brownish-grey bark; Leaves opposite, petioled, orbicular, cordate, acuminate, coriaceous, glabrescent or tomentose; Flowers yellow, downy, in globose, peduncled heads; Capsules of 2 dehiscent cocci, dry, many seeded.

Flowering and Fruiting: June-January

10. Scientific Name: *Aegle marmelos* (L.) Correa

Synonym: *Bilacus marmelos* (L.) Kuntze

Family: Rutaceae

Common Name: Wood Apple

Vernacular Name: Bel

Habitat: Open tropical areas

Uses: Marmalade, Confectionery, Snake bite antidote, Medicinal

IUCN Threat listing: Not Assessed (NA)

Description: Small to medium, glabrous, deciduous tree; Stem up to 10 m high, bark corky grey, branches ascending; Leaves alternate, trifoliate, axillary spines sharp, hard and straight, leaflets ovate or rhomboid, acuminate, crenate serrate, gland-dotted; Flowers in panicles, greenish-white, sweet-scented; Fruit a globose drupe, 5-15 cm across, with woody shell and sweet pulp.

Flowering and Fruiting: March-May

11. Scientific Name: *Aeschynomene americana* L.

Synonym: *Hippocrepis mimulosa* Noronha

Family: Leguminosae

Common Name: American joint vetch

Vernacular Name: Sol Ghas

Habitat: Open tropical grasslands

Uses: Fodder

IUCN Threat listing: Not Assessed

Description: An erect or ascending, rarely almost prostrate, annual or short-lived perennial shrub; Leaves pinnate, grey green to dark green, often with purplish tinge, pinnae 8-38 pairs; Flowers few, in branched racemes, corolla pinkish to pale mauve or yellow-orange; Pods mostly slightly curved, of 3-9 segments.

Flowering and Fruiting: July-November

12. Scientific Name: *Aeschynomene indica* L.

Synonym: *Smithia aspera* Roxb.

Family: Leguminosae

Common Name: Indian joint vetch

Vernacular Name: Phulan

Habitat: Open tropical grasslands

Uses: Green manure, Fodder

IUCN Threat listing: Least Concerned (LC)

Description: Annual, hardy, aquatic herb; Stem glabrous, pale-green, with many terete branches; Leaves 5-10 cm long, paripinnate, leaflets 40-60, oblong-lanceolate to elliptic; Flowers 1-4, in axillary racemes, yellow, veined with purple. Pods muricate on both surfaces, papillose, 6-10-jointed, dotted with black.

Flowering and Fruiting: August-December

13. Scientific Name: *Afrohybanthus enneaspermus* (L.) Flicker

Synonym: *Hybanthus enneaspermus* (L.) F. Müll.

Family: Violaceae

Common Name: Spade flower

Vernacular Name: Ratan Purush

Habitat: Open humid tropical areas

Uses: Medicinal

IUCN Threat listing: Not Assessed (NA)

Description: A perennial herb or a small shrub, up to 60 cm high; Leaves linear to lance like, 1-5 cm long; Flowers solitary, pink-purple, spade shaped, anthers without appendages; Capsules 4-9 mm long, seeds 5-10.

Flowering and Fruiting: July-September

14. Scientific Name: *Agave sisalana* Perrine

Synonym: *Agave amaniensis* Trel. & Nowell

Family: Asparagaceae

Common Name: Sisal hemp

Vernacular Name: Sisal

Habitat: Open tropical areas

Uses: Fibre, Medicinal

IUCN Threat listing: Not Assessed (NA)

Description: Perennial, stout, scapigerous shrub; Trunk prostrate or ascending; Leaves stout, more or less glaucous, spinescent-dentate, marginal spines pointing downwards; Fascicles of flowers crowded at the end of ascending branches, scape 5-8 m high.

Flowering and Fruiting: February-October

15. Scientific Name: *Ageratum conyzoides* L.

Synonym: *Eupatorium conyzoides* (L.) E. H. Krause

Family: Asteraceae

Common Name: Goat weed

Vernacular Name: Kubbi

Habitat: Open tropical grasslands

Uses: Medicinal, Insect repellent

IUCN Threat listing: Not Assessed (NA)

Description: Softly hairy, erect annual; Stems branched, terete, 30-90 cm high; Leaves opposite, sub-acute, petioled, ovate, crenate, cuneate to sub-cordate at base; Flowers tubular, pale purplish-blue to white arranged in numerous, discoid, dense, terminal corymbose heads; Achenes angled, glandular, black.

Flowering and Fruiting: August-December.

16. Scientific Name: *Albizia chinensis* (Osbeck) Merr.

Synonym: *Acacia stipulata* DC.

Family: Leguminosae

Common Name: Chinese Albizia

Vernacular Name: Kala Siris

Habitat: Open tropical areas

Uses: Fuel wood, Medicinal

IUCN Threat listing: Not Assessed (NA)

Description: Medium-sized, unarmed, deciduous or evergreen tree, bark thin, smooth, lenticellate, dark gray; Leaves bipinnate, with a stout, lenticellate, glabrescent rachis, about 10-25 cm long; Peduncles in clusters of up to 5, 1-3 cm long, glomerule of 10-20 pentamerous, dimorphic flowers; Fruit an indehiscent pod, thin, flat, strap-shaped.

Flowering and Fruiting: November-July

17. Scientific Name: *Albizia lebbeck* (L.) Benth.

Synonym: *Acacia macrophylla* Bunge

Family: Leguminosae

Common Name: Lebbeck

Vernacular Name: Siris

Habitat: Open tropical areas

Uses: Fuel wood, Shade tree, Timber, Medicinal

IUCN Threat listing: Not Assessed (NA)

Description: Large, unarmed, deciduous tree; Crown spreading, bark pale, irregularly cracked; Leaves bipinnate, rachis stout, pinnae 2-3 pairs, leaflets 4-9 pairs, oblong, obtuse, oblique, sub-sessile, pale green; Flowers in heads, white, fragrant, peduncles 5-10 cm, forming short corymbose racemes; Pods 10-30 cm long, thin, flat, straw-coloured, 6-10-seeded.

Flowering and Fruiting: March-December

18. Scientific Name: *Albzia odoratissima* (L. *f.*) Benth.

Synonym: *Mimosa odoratissima* L. *f.*

Family: Leguminosae

Common Name: Ceylon rosewood

Vernacular Name: Siris

Habitat: Open sunny landscapes

Uses: Fuel wood, avenue

IUCN Threat listing: Not Assessed (NA)

Description: A large, semi-deciduous or semi-evergreen tree with spreading crown; Leaves abruptly 2-pinnate, pinnae 6-8, leaflets 16-40; Flowers white, fragrant, sessile, in numerous small heads; Pods thin, flexible, 15-20 cm long, brown.

Flowering and Fruiting: May-July

19. Scientific Name: *Albizia procera* (Roxb.) Benth.

Synonym: *Mimosa procera* Roxb.

Family: Leguminosae

Common Name: Tall Albizia

Vernacular Name: Lal Siris

Habitat: Open tropical areas

Uses: Fuel wood, Shade tree, Timber, Medicinal

IUCN Threat listing: Not Assessed (NA)

Description: Large tree; Stem with greenish-white, smooth bark, exfoliating in thin flakes; Leaves 25-35 cm long, pinnae 3-5 pairs, leaflets 12-24 on each pinnae; Flowers whitish-yellow, in small, 15-20-flowered heads, arranged in clusters of 2-5 on 30-60 cm long racemes; Pods 10-15 cm long, smooth, brown, 8-12-seeded.

Flowering and Fruiting: July-December

20. Scientific Name: *Aloe vera* (L.) Burm. *f.*

Synonym: *Aloe barbadensis* Mill.

Family: Xanthorrhoeaceae

Common Name: Aloe

Vernacular Name: Ghritkumari

Habitat: Open tropical grasslands

Uses: Cosmetic, Burns, Medicinal

IUCN Threat listing: Not Assessed (NA)

Description: Acaulescent, stoloniferous perennial, up to 60 cm; Leaves in rosettes near ground, dense, thick, soft, glaucous-green with purplish tinge, lanceolate, long acuminate bearing thorns at the edges; Flowers on a long, usually simple scape, drooping, dull red, in simple, erect racemes, perianth reddish-yellow, cylindric.

Flowering and Fruiting: November-February

21. Scientific Name: *Alstonia scholaris* (L.) R. Br.

Synonym: *Alstonia kurzii* Hook. *f.*

Family: Apocyanaceae

Common Name: Devil tree

Vernacular Name: Kuraya

Habitat: Open humid tropical areas

Uses: Crockery items, Snake Bite, Medicinal

IUCN Threat listing: Least Concerned (LC)

Description: Medium to large tree, up to 20 m tall, bark pale brown, smooth, scaly, with large horizontal lenticels, peeling off in rectangular flakes; Leaves whorled, glossy dark green above, pale grey-green below, elliptic to narrowly obovate, sub-coriaceous or leathery; Flowers in dense terminal cymes, white, creamy-greenish white, yellow or cream, petals partially-pubescent; Fruits slender, linear, dehiscent follicles, 20-40 cm long.

Flowering and Fruiting: October-February

22. Scientific Name: *Alternanthera brasiliana* (L.) Kuntze

Synonym: *Gomphrena brasiliana* L.

Family: Amaranthaceae

Common Name: Purple joyweed

Vernacular Name: Lal Patti

Habitat: Open tropical grasslands

Uses: Edible, Medicinal

IUCN Threat listing: Not Assessed (NA)

Description: Perennial herb up to 90 cm tall; Leaves elliptic, mostly vibrant pink, purplish-red near the centre; Flowers small, white, produced in the leaf axils as globular heads.

Flowering and Fruiting: July-November

23. Scientific Name: *Alternanthera ficoidea* (L.) R. Br.

Synonym: *Gomphrena ficoidea* L.

Family: Amaranthaceae

Common Name: Sanguinarea

Vernacular Name: Gudrisag

Habitat: Open tropical grasslands

Uses: Edible, Medicinal

IUCN Threat listing: Not Assessed (NA)

Description: Prostrate, annual herb; Stem diffusely branching; Leaves fleshy, up to 5 cm, linear-lanceolate, or elliptic-oblong, hairy beneath, apex obtuse, base acute or cuneate; Flowers sessile, white and shining, arranged in small, axillary, sessile heads; Fruit an urticle, broadly obcordate.

Flowering and Fruiting: June-August

24. Scientific Name: *Alysicarpus monilifer* (L.)DC.

Synonym: *Hedysarum monilifer* L.

Family: Leguminosae

Common Name: Necklace pod

Vernacular Name: Chauli

Habitat: Open tropical grasslands

Uses: Fodder

IUCN Threat listing: Not Assessed (NA)

Description: Prostrate, procumbent, perennial herb; Stem densely tufted, much branched, clothed with spreading hairs; Leaves simple, oblong, obtuse, often cordate, shortly-petioled, apex obtuse, base cordate; Flowers 4-10, shortly stalked in erect, pedunculate racemes, pink; Pods moniliform, 2-6-jointed, segments sub-globose, densely clothed with short, hooked hairs.

Flowering and Fruiting: November-January

25. Scientific Name: *Amaranthus hybridus* L.

Synonym: *Amaranthus cruentus* L.

Family: Amaranthaceae

Common Name: Green Amaranth

Vernacular Name: Chaulai

Habitat: Open tropical grasslands

Uses: Edible, Medicinal

IUCN Threat listing: Not Assessed (NA)

Description: An erect, annual herbaceous plant, 10 cm to 2 m in height, stems thick, often ribbed or tinged with red; Leaves alternate, long-stalked, ovate to rhombic-ovate; Flowers numerous, green, crowded into finger-like spikes, forming long, dense, terminal panicle, with axillary spikes below; Seeds round, dark brown, shiny, somewhat flattened.

Flowering and Fruiting: July-September

26. Scientific Name: *Amaranthus spinosus* L.

Synonym: *Galliaria spinosa* (L.) Nieuwl.

Family: Amaranthaceae

Common Name: Spiny Amaranth

Vernacular Name: Kantili Chaulai

Habitat: Open tropical grasslands

Uses: Edible, Medicinal

IUCN Threat listing: Not Assessed (NA)

Description: Annual spinous herb, up to 1.5 m high; Stem often tinged with red; Leaves ovate-lanceolate, sub-entire, glabrous, apex acute, base cuneate; Flowers sessile, lower ones in axillary, dense clusters, upper ones in spikes or panicles, greenish-yellow; Fruit capsule, ovoid, rugose.

Flowering and Fruiting: June-December

27. Scientific Name: *Ampelocissus latifolia* (Roxb.) Planch.

Synonym: *Vitis latifolia* Roxb.

Family: Vitaceae

Common Name: Wild Grape

Vernacular Name: Ban Angur

Habitat: Open humid tropical areas

Uses: Edible, Medicinal

IUCN Threat listing: Not Assessed (NA)

Description: Large, twining shrub: Stem soft, porous, furrowed, branches glabrous; Leaves simple, long petiolate, ovate, 5-lobed, tendril slender and forked; Flowers yellow or reddish-brown, in compound panicles or umbellate cymes; Fruit a berry, globose, black when ripe.

Flowering and Fruiting: July-September

28. Scientific Name: *Andrographis paniculata* (Burm. *f.*) Nees

Synonym: *Justicia paniculata* Burm. *f.*

Family: Acanthaceae

Common Name: Creat

Vernacular Name: Kalmegh

Habitat: Open tropical grasslands

Uses: Medicinal

IUCN Threat listing: Not Assessed (NA)

Description: Erect, glabrous, annual herb; Stem quadrangular, winged; Leaves shortly-petioled, ovate-lanceolate; Flowers small, white, spotted with rose-purple, borne in lax, axillary and terminal, unilateral racemes; Fruit capsule, about 2 cm long, tapering at both ends.

Flowering and Fruiting: November-January

29. Scientific Name: *Anisomeles indica* (L.) Kuntze

Synonym: *Nepeta indica* L.

Family: Lamiaceae

Common Name: Catmint

Vernacular Name: Kala Bhangra

Habitat: Open tropical areas

Uses: Aromatic, Medicinal

IUCN Threat listing: Not Assessed (NA)

Description: Erect, pubescent herb or undershrub; Stem 4-angled, up to 1.5 m; Leaves petioled, ovate, crenate-serrate, pubescent on both sides, petioles hairy; Flowers in axillary whorls, forming interrupted terminal spikes, pale purple to pink; Nutlets 4, blackish, smooth and shining.

Flowering and Fruiting: January-April

30. Scientific Name: *Annona squamosa* L.

Synonym: *Annona forskahlii* DC.

Family: Annonaceae

Common Name: Sugar Apple

Vernacular Name: Sitaphal

Habitat: Open humid tropical areas

Uses: Edible, Insect repellent, Medicinal

IUCN Threat listing: Not Assessed (NA)

Description: A small, deciduous shrub, 3-6 m tall, branches irregularly spreading, densely pubescent when young; Leaves alternate, ovate-oblong or elliptic-oblong, thin, sparsely downy, dark green above; Flowers small, pendulous, occur singly or in pairs in the leaf axils of young shoots or opposite the leaves; Fruit a syncarp, by fusion of numerous ovaries, irregularly heart-shaped, 5-20 cm in diameter.

Flowering and Fruiting: April-November

31. Scientific Name: *Anogeissus latifolia* (Roxb. *ex* DC.) Bedd.

Synonym: *Andersonia altissima* Roxb. *ex* Wight & Arn.

Family: Combretaceae

Common Name: Axle wood tree

Vernacular Name: Dhawa

Habitat: Humid tropical areas

Uses: Fuelwood, Gum, Medicinal

IUCN Threat listing: Not Assessed (NA)

Description: Large tree; Stem with smooth white-grey bark; Leaves alternate or sub-opposite, ovate-lanceolate, apex obtuse or emarginate, base rounded; Flowers sessile, small, in globose heads; Fruit a dry drupe, nearly orbicular, yellowish-brown, many in a globular head.

Flowering and Fruiting: April-December

32. Scientific Name: *Argemone mexicana* L.

Synonym: *Echtrus mexicanus* (L.) Nieuwland

Family: Papaveraceae

Common Name: Mexican Poppy

Vernacular Name: Pili

Habitat: Open tropical grasslands

Uses: Medicinal, Worm repellent

IUCN Threat listing: Not Assessed (NA)

Description: Erect thistle-like perennial herb; Stems glaucescent, prickly, up to 90 cm, with yellow latex; Leaves 7-15 cm long, upper ones sessile, with semi-amplexicaul base, lower ones petioled in a rosette, sinuate, pinnatifid; Flowers terminal or solitary, shortly-pedicelled; Fruit a capsule, oblong-ellipsoid, 3-6 valved, prickly.

Flowering and Fruiting: Throughout the year

33. Scientific Name: *Argyreia nervosa* (Burm. *f.*) Bojer

Synonym: *Convolvulus nervosus* Burm. *f.*

Family: Convolvulaceae

Common Name: Elephant creeper

Vernacular Name: Ghav Bel

Habitat: Open tropical areas

Uses: Wound healing, Medicinal

IUCN Threat listing: Not Assessed (NA)

Description: Creeper, woody at base, densely white pubescent when young, glabrescent; Leaves cordate, 18-27 cm long, apically acute to obtuse; Flowers in cymes, on long, white, tomentose peduncles, bracts ovate-lanceolate, subtending; Fruit indehiscent, baccate, dry, subglobose, 1-1.5 cm long.

Flowering and Fruiting: December-June

34. Scientific Name: *Arisaema tortuosum* (Wall.) Schott

Synonym: *Arum tortuosum* Wall.

Family: Araceae

Common Name: Whipcord Cobra Lily

Vernacular Name: Bagh Jandra

Habitat: Open humid tropical areas

Uses: Insecticidal, Ectoparasite of cattle

IUCN Threat listing: Not Assessed (NA)

Description: Perennial rhizomatous herb; Stem a robust rhizome; Leaves dark green, lower ones sheathing, forming pseudostem, 5-7-foliolate, leaflets oblong, elliptic; Flowers in spadix, on 5-10 cm peduncle, spathe green; Fruit reddish berries.

Flowering and Fruiting: April-July

35. Scientific Name: *Arthraxon hispidus* (Thunb.) Makino

Synonym: *Arthraxon hispidus* (Thunb.) Merr.

Family: Poaceae

Common Name: Hairy joint grass

Vernacular Name: Jal Chara

Habitat: Open tropical grasslands

Uses: Fodder

IUCN Threat listing: Not Assessed (NA)

Description: A perennial, herbaceous grass, usually gregarious in moist areas; Leaves relatively short and broad, narrowly obovate, up to 5 cm long; Spikelets sessile, up to 7 mm long, lower glume lanceolate, convex, with scabrid veins.

Flowering and Fruiting: September-March

36. Scientific Name: *Artocarpus heterophyllus* Lam.

Synonym: *Artocarpus jaca* Lam.

Family: Moraceae

Common Name: Jackfruit

Vernacular Name: Kathal

Habitat: Open humid tropical areas

Uses: Edible, Aromatic, Furniture

IUCN Threat listing: Not Assessed (NA)

Description: An evergreen tree, growing up to height of 20 m, trunk usually tubercled, bark blackish-grey, lenticels reddish-brown, irregularly flaky; Leaves simple, alternate, spiral, stipules oblong or lanceolate; Flowers unisexual, in spikes, enclosed by spathe-like bracts, male flowers in narrow, cylindrical-oblong catkin on young branches, female flowers usually cauliflorus.

Flowering and Fruiting: December-June

37. Scientific Name: *Asparagus racemosa* Willd.

Synonym: *Protasparagus racemosus* (Willd.) Oberm

Family: Asparagaceae

Common Name: Asparagus

Vernacular Name: Jogilati

Habitat: Humid tropical areas

Uses: Medicinal

IUCN Threat listing: Not Assessed (NA)

Description: Scandent, much-branched, tall, undershrub, roots tuberous; Branches grooved, rough, cladodes in tufts of 2-6, spreading; Flowers white, in dense racemes, shortly-pedicelled; Berry globose, reddish or violet.

Flowering and Fruiting: October-January

38. Scientific Name: *Averrhoa carambola* L.

Synonym: *Averrhoa pentandra* Blanco

Family: Oxalidaceae

Common Name: Star fruit

Vernacular Name: Karmal

Habitat: Humid tropical areas

Uses: Edible, Medicinal

IUCN Threat listing: Not Assessed (NA)

Description: A small, evergreen tree, up to 12 m high, densely branched, young parts finely pubescent or glabrous; Leaves compound, 7-25 cm, leaflets 3-13, blades ovate to elliptic; Inflorescences axillary or rameal panicles or cymes, branches and flower buds crimson, flowers numerous, small; Berries yellow to yellow-brown, oblong, deeply 3-6-ribbed, stellate in cross section, very fleshy.

Flowering and Fruiting: April-November

39. Scientific Name: *Azadirachta indica* A. Juss.

Synonym: *Melia azadirachta* L.

Family: Meliaceae

Common Name: Indian Lilac

Vernacular Name: Neem

Habitat: Open tropical areas

Uses: Medicinal, Insect repellent, Edible

IUCN Threat listing: Not Assessed (NA)

Description: Large attractive tree; Trunk massive, branches spreading; Leaves imparipinnate, crowded at the end of branches, leaflets 10-12 per leaf, ovate-lanceolate, sub-opposite, bright green and shining; Flowers white, scented, in dense axillary panicles; Drupes ovoid, smooth, green, turning pale yellow on ripening.

Flowering and Fruiting: January-August

40. Scientific Name: *Baccharoides anthelmintica* (L.) Moench

Synonym: *Vernonia anthelmintica* (L.) Willd.

Family: Asteraceae

Common Name: Ironweed

Vernacular Name: Kali Jeeri

Habitat: Open tropical areas

Uses: Seasoning agent, Medicinal

IUCN Threat listing: Not Assessed (NA)

Description: Tall, robust, glandular-pubescent annual; Stem much branched, 1.2-1.8 m high; Leaves rough, ovate-lanceolate, shortly petioled, acute, coarsely serrate, rough, long pointed; Flowers pink, heads many-flowered, sub-corymbose, involucral bracts lanceolate, narrow, acute; Achenes 10-ribbed, black, hairy, pappus white.

Flowering and Fruiting: March-September

41. Scientific Name: *Bambusa vulgaris* Schrad. *ex* J.C. Wendl.

Synonym: *Bambusa vulgaris Schrad. ex* J. C. Wendl. var. *vittata* Rivière

Family: Poaceae

Common Name: Golden Bamboo

Vernacular Name: Baans

Habitat: Humid tropical areas

Uses: Handicrafts, Poles, Erosion Control, Ornamental

IUCN Threat listing: Not Assessed (NA)

Description: A robust herb forming moderate to loose clumps, culms lemon-yellow, with green stripes; Leaves deciduous, with dense pubescence, blades narrowly lanceolate; Flowers bracteate, in large indeterminate inflorescences, forming pseudo-spikelets, florets with 3 lodicules.

Flowering and Fruiting: Monocarpic

42. Scientific Name: *Barleria prattensis* Santapau

Synonym: *Barleria gibsonii* Dalz.

Family: Acanthaceae

Common Name: Porcupine flower

Vernacular Name: Jhinti

Habitat: Open humid tropical areas

Uses: Ornamental

IUCN Threat listing: Not Assessed (NA)

Description: Herbaceous perennial; Leaves opposite, ovate or elliptic, acuminate, apex mucronate, base acute or attenuate, membranous, glabrous; Flowers solitary, axillary, terminating in short racemes, bracts linear-subulate, corolla lobes sub-equal, broadly ovate, pink-purple; Capsules nearly 2 cm long, oblong, seeds 4, orbicular, glabrous.

Flowering and Fruiting: June-October

43. Scientific Name: *Bauhinia divaricata* L.

Synonym: *Bauhinia retusa* Poir.

Family: Leguminosae

Common Name: Pom-pom orchid tree

Vernacular Name: Tawar

Habitat: Open tropical areas

Uses: Avenue tree, Fuel wood

IUCN Threat listing: Not Assessed (NA)

Description: A mid-size tree, with spreading crown, stems woody, twigs hairy or smooth; Leaves alternate, blade sub-orbicular to lanceolate; Inflorescence in terminal or sub-terminal clusters, short and dense, with 20-50 flowers; Fruit a legume, linear-oblong, 8 to 12 cm long.

Flowering and Fruiting: September-March

44. Scientific Name: *Bauhinia purpurea* L.

Synonym: *Phanera purpurea* (L.) Benth.

Family: Leguminosae

Common Name: Purple orchid tree

Vernacular Name: Koilar

Habitat: Open humid tropical areas

Uses: Edible, Fodder, Medicinal, Erosion control

IUCN Threat listing: Least Concerned (LC)

Description: Small/moderate-sized tree; Bark dark brown; Leaves longer than broad, ovate-cordate, coriaceous, cleft nearly half-way down; Flowers rose-coloured, in corymbose or paniculate racemes; Pods 15-30 cm long, flat, oblong, ellipsoid.

Flowering and Fruiting: October-March

45. Scientific Name: *Bauhinia racemosa* Lam.

Synonym: *Bauhinia parviflora* Vahl

Family: Leguminosae

Common Name: Bidi leaf tree

Vernacular Name: Kachnar

Habitat: Open humid tropical areas

Uses: Medicinal, Cigarette Wrapping

IUCN Threat listing: Not Assessed (NA)

Description: Small deciduous tree; Bark dark-brownish, branches crooked, main trunk with vertical cracks; Leaves broader than long, deeply cleft, ovate, apex acute, base cordate; Flowers in axillary racemes or panicles; Pods stalked, 15-30 cm, falcate, torulose, indehiscent.

Flowering and Fruiting: March-December

46. Scientific Name: *Bauhinia vahlii* Wight & Arn.

Synonym: *Phanera vahlii* (Wight & Arn.) Benth.

Family: Leguminosae

Common Name: Maloo creeper

Vernacular Name: Mohlain

Habitat: Open humid tropical areas

Uses: Edible, Medicinal

IUCN Threat listing: Not Assessed (NA)

Description: Profuse evergreen climber; Branches thick, cable-like; Leaves ca. 45 cm across, ± as broad as long, deeply cordate, dark green; Flowers off white, in dense, terminal long-peduncled corymbs or corymbose racemes; Pods flat, woody, clothed with dense rusty tomentum.

Flowering and Fruiting: July-November

47. Scientific Name: *Begonia picta* Sm.

Synonym: *Begonia echinata* Royle

Family: Begoniaceae

Common Name: Painted leaf *Begonia*

Vernacular Name: Patharchata

Habitat: Humid tropical areas

Uses: Edible, Medicinal, Mordant

IUCN Threat listing: Not Assessed (NA)

Description: Annual, succulent, slender herb; Stem suffused with reddish, glabrous; Leaves one to many, long-petioled, ovate-cordate, irregularly dentate, apex acuminate; Flowers male and female borne separately, pink or pale rose, in dichasial cymes; Capsules trigonous, winged, hairy.

Flowering and Fruiting: August-October

48. Scientific Name: *Bidens pilosa* L.

Synonym: *Kerneria pilosa* (L.) Lowe

Family: Asteraceae

Common Name: Blackjack

Vernacular Name: Kumur

Habitat: Open humid tropical areas

Uses: Edible, Medicinal

IUCN Threat listing: Not Assessed (NA)

Description: Erect herb, 30-100 cm tall; Stem robust, glabrous near the base, pubescent upwards; Leaves variable, simple or compound, when compound, leaflets ovate, toothed; Flowers yellow, in heads arranged in corymbose panicles; Achenes ribbed, glabrous, black.

Flowering and Fruiting: April-November

49. Scientific Name: *Biophytum sensitivum* (L.) DC.

Synonym: *Oxalis sensitiva* L.

Family: Oxalidaceae

Common Name: Little tree plant

Vernacular Name: Laajvanti

Habitat: Humid tropical areas

Uses: Ornamental, Medicinal

IUCN Threat listing: Not Assessed (NA)

Description: Annual herb, up to 15 cm; Leaves 2-8 cm, leaflets 7-15 pairs, terminal pair largest, obovate; Umbels 5-10, 7-10 flowered, flowers yellow, with brick red tinge inside corolla; Capsules glandular-pubescent along septa, seeds 2-4 in each locule.

Flowering and Fruiting: September-December

50. Scientific Name: *Blainvillea acmella* (L.) Philipson

Synonym: *Verbesina acmella* L.

Family: Asteraceae

Common Name: Para Cress Flower

Vernacular Name: Kanghi

Habitat: Open tropical grasslands

Uses: Medicinal, Dye

IUCN Threat listing: Not Assessed (NA)

Description: Erect, stiff, roughly pubescent herb; Stem up to 60 cm high; Leaves petioled, lower opposite, upper alternate, ovate-lanceolate, acute or acuminate, crenate-serrate; Flowers in radiate heads, white or yellow; Achenes curved, angled, pappus minutely bristly.

Flowering and Fruiting: April-November

51. Scientific Name: *Boerhavia diffusa* L.

Synonym: *Boerhavia surinamensis* Miq.

Family: Nyctaginaceae

Common Name: Red Spiderling

Vernacular Name: Punarnava

Habitat: Open humid tropical areas

Uses: Edible, Medicinal, Fodder

IUCN Threat listing: Not Assessed (NA)

Description: Perennial, decumbent herb; Stem sub-erect to prostrate; Leaves petiolate, in unequal pairs, ovate-sub-orbicular, glabrous, rounded-cordate at base; Flowers crowded in bracteate, long-stalked umbels, pale pink; Fruit clavate or obovoid, 5-ribbed, glandular.

Flowering and Fruiting: August-December

52. Scientific Name: *Bombax ceiba* L.

Synonym: *Salmalia malabarica* (DC.) Schott & Endl.

Family: Malvaceae

Common Name: Cotton tree

Vernacular Name: Semal

Habitat: Humid tropical areas

Uses: Fuel wood, Fibre

IUCN Threat listing: Not Assessed (NA)

Description: Large, deciduous tree; Stem huge, straight, covered with very hard, large, conical prickles when young, branches spreading; Leaves digitate, leaflets lanceolate, acuminate, entire; Flowers in clusters, fleshy, scarlet; Capsules oblong-ovoid, 5-angled, containing smooth seeds, surrounded by long, white silky hairs.

Flowering and Fruiting: January-April

53. Scientific Name: *Bonnaya ciliata* (Colsm.) Spreng

Synonym: *Lindernia ciliata* (Colsm.)Pennell

Family: Linderniaceae

Common Name: Hairy Slitwort

Vernacular Name: Kankare Jhar

Habitat: Open humid tropical areas

Uses: Fodder, Medicinal

IUCN Threat listing: Not Assessed (NA)

Description: Annual herb, of moist and open areas; Leaves elliptic-oblong, base rounded, margin aristate-dentate, apex obtuse or acute; Flowers in lax terminal racemes, pedicels 1-2 mm, bracts leafy, corolla pink or bluish-pink; Capsule 1-1.5 cm long, linear, septum winged, exceeding the fruiting calyx.

Flowering and Fruiting: June-October

54. Scientific Name: *Bougainvillea spectabilis* Willd.

Synonym: *Tricycla spectabilis* Poir.

Family: Nyctaginaceae

Common Name: Bougainvillea

Vernacular Name: Buganbel

Habitat: Open tropical areas

Uses: Ornamental, Medicinal

IUCN Threat listing: Not Assessed (NA)

Description: A perennial, woody climber, 7-12 m long, commonly cultivated; Leaves petiolate, blade round-ovate to suborbicular, sometimes blotched with white; Pseudanthia 3-flowered, peduncle 6-17 mm long, slender, bracts pink or purplish-red, ovate or broadly ovate, abruptly acute, acuminate, or sometimes obtuse at apex.

Flowering and Fruiting: March-July

55. Scientific Name: *Brassica rapa* (DC.) Metzg.

Synonym: *Brassica campestris* L.

Family: Brassicaceae

Common Name: Field Mustard

Vernacular Name: Rayi

Habitat: Open tropical areas

Uses: Edible, Oil yielding, Medicinal

IUCN Threat listing: Not Assessed (NA)

Description: Herbaceous, annual or biennial, 30-120 cm high branched; Leaves large, soft, smooth or soft-hairy, up to 50 cm long, pinnatifid (deeply lobed) or lyrate; Flowers in racemes, yellow, four-parted and cross-shaped; Fruit is capsular, splits open when mature.

Flowering and Fruiting: April-August

56. Scientific Name: *Bridelia retusa* (L.) A. Juss.

Synonym: *Clutia retusa* L.

Family: Phyllanthaceae

Common Name: Spinous Kino Tree

Vernacular Name: Kasai

Habitat: Open humid tropical areas

Uses: Edible, Medicinal, Fuel wood

IUCN Threat listing: Not Assessed (NA)

Description: Moderate-sized tree; Stem up to 6 m high, with strong spines; Leaves alternate, elliptic-oblong to obovate, greenish to reddish, crenate or entire; Flowers dioecious, yellow, small, sub-sessile; Drupes sub-globose, seated on a persistent calyx.

Flowering and Fruiting: July-September

57. Scientific Name: *Buchanania cochinchinensis* (Lour.) Almeida

Synonym: *Buchanania lanzan* Spreng.

Family: Anacardiaceae

Common Name: Almondette

Vernacular Name: Chaar

Habitat: Open humid tropical areas

Uses: Edible, Medicinal

IUCN Threat listing: Not Assessed (NA)

Description: A medium size, to large evergreen tree; Leaves 15-25 cm long, thickly coriaceous, broadly oblong; Flowers small, sessile, greenish white, in terminal and axillary pyramidal panicles; Fruits drupaceous, obliquely lentiform, black.

Flowering and Fruiting: January-May

58. Scientific Name: *Butea monosperma* (Lam.) Taub.

Synonym: *Erythrina monosperma* Lam.

Family: Leguminosae

Common Name: Flame of the forest

Vernacular Name: Chula

Habitat: Open humid tropical areas

Uses: Fodder, Timber, Fuel wood, Resin, Dye, Medicinal

IUCN Threat listing: Not Assessed (NA)

Description: Moderate-sized deciduous tree; Trunk crooked with greyish, corky bark; Leaves trifoliate, leaflets hard, coriaceous, rhomboid-orbicular, base cuneate or deltoid; Flowers bright orange-red in colour, appear before leaves, in axillary and terminal racemes; Pods pendulous, stalked, silky tomentose.

Flowering and Fruiting: March-September

59. Scientific Name: *Caesalpinia bonduc* (L.) Roxb.

Synonym: *Caesalpinia bonducella* (L.) Fleming

Family: Leguminosae

Common Name: Fever nut

Vernacular Name: Gattara

Habitat: Open tropical areas

Uses: Medicinal

IUCN Threat listing: Not Assessed (NA)

Description: Densely prickled shrub; Stem much branched; Leaves paripinnate leaves, pinnae 6-8 pairs, leaflets opposite, rachis densely prickled; Flowers in axillary, simple or panicled racemes, denser towards the top; Pods oblong, densely armed with sharp, wiry prickles on the faces.

Flowering and Fruiting: March-October

60. Scientific Name: *Caesalpinia decapetala* (Roth) Alston

Synonym: *Reichardia decapetala* Roth

Family: Leguminosae

Common Name: Mysore thorn

Vernacular Name: Kirkich

Habitat: Open tropical areas

Uses: Hedge Plant, Ornamental, Medicinal, Insecticidal

IUCN Threat listing: Not Assessed (NA)

Description: A robust, sprawling shrub or climber, stem with numerous straight to hooked thorns; Leaves bipinnately compound, dark green above, paler beneath, up to 30 cm, pinnae 3-15 pairs, of 10-24 leaflets, pinnules elliptic-oblong to ovate, rounded at the apex; Flowers pale yellow, borne in axillary and terminal racemes, 10-40 cm long; Fruits dehiscent pods, 6-11 cm long, seeds 4-9, black, ellipsoid, flattened.

Flowering and Fruiting: Throughout the Year

61. Scientific Name: *Caesalpinia pulcherrima* (L.) Sw.

Synonym: *Poinciana pulcherrima* L.

Family: Leguminosae

Common Name: Peacock flower

Vernacular Name: Guletura

Habitat: Open humid tropical areas

Uses: Medicinal, Antimicrobial, Ornamental

IUCN Threat listing: Not Assessed (NA)

Description: A shrub, 3-4 m high, branches delicate with scattered thorns; Leaves alternate, bipinnate, pinnae 4-8 pairs, each with 7-11 pairs of elliptic, obtuse, obliquely inequilateral, light green leaflets; Flowers in long terminal racemes, red-and-yellow, petals crinkly-edged, stamens with long filaments of the colour of petals; Fruit a coriaceous pod, oblong, smooth, brown to black, several-seeded.

Flowering and Fruiting: Throughout the Year

62. Scientific Name: *Caesulia axillaris* Roxb.

Synonym: *Caesulia lancifolia* Sw.

Family: Asteraceae

Common Name: Pink node flower

Vernacular Name: Thuk Jhar

Habitat: Humid tropical areas

Uses: Edible, Medicinal, Antimicrobial

IUCN Threat listing: Least Concerned (LC)

Description: An erect, annual herb, of moist areas, stem stout, pink, succulent, swollen at nodes; Leaves alternate, sessile, lanceolate, 4-15 cm long, pointed, distantly toothed, narrow at base; Capitulae stalkless, 1-2 cm across, axillary, flowers white, tubular, with 5 linear petals.

Flowering and Fruiting: September-November

63. Scientific Name: *Cajanus cajan* (L.) Millsp.

Synonym: *Cajan cajan* (L.) Millsp.

Family: Leguminosae

Common Name: Pigeonpea

Vernacular Name: Arhar

Habitat: Open tropical areas

Uses: Edible, Medicinal, Ornamental

IUCN Threat listing: Not Assessed (NA)

Description: An erect, perennial shrub, up to 4 m, stems and branches pubescent; Leaves pinnately trifoliolate, leaflets 2.5-10 cm long, 1.5-3.5 cm broad, elliptic to lanceolate, acute, velvety pubescent above, pilose beneath; Flowers on short racemes, calyx velvety pubescent and glandular, corolla bright yellow with reddish brown or crimson lines; Pods 5-10 cm long, pubescent and glandular, seeds 5-10.

Flowering and Fruiting: April-December

64. Scientific Name: *Cajanus scarabaeoides* (L.)Thouars

Synonym: *Atylosia scarabaeoides* (L.)Benth.

Family: Leguminosae

Common Name: Showy Pigeonpea

Vernacular Name: Ban Rahar

Habitat: Open tropical grasslands

Uses: Fodder, Medicinal

IUCN Threat listing: Least Concerned (LC)

Description: Slender twiner, sometimes woody at base; Stem densely grey-pubescent; Leaves trifoliate, leaflets oblong or elliptic-obovate, grey-pubescent above, densely downy beneath; Flowers 2-6, on short, densely pubescent, axillary peduncle; Pods 2.0-2.5 cm, straight, densely clothed with silky brownish hairs.

Flowering and Fruiting: October-December

65. Scientific Name: *Caladium bicolor* (Aiton) Vent.

Synonym: *Arum bicolor* Aiton

Family: Araceae

Common Name: Angel wings

Vernacular Name: Lal Kochai

Habitat: Humid tropical areas

Uses: Ornamental, Medicinal

IUCN Threat listing: Not Assessed (NA)

Description: A perennial, clump forming herb, up to 80 cm high; Leaves large, thin, arrowhead shaped, with a variety of patterns and colors; Flowers small, inconspicuous, covered by a small spathe.

Flowering and Fruiting: July-August

66. Scientific Name: *Calliandra haematocephala* Hassk.

Synonym: *Anneslia haematocephala* (Hassk.) Britton & P.Wilson

Family: Leguminosae

Common Name: Red Powder Puff

Habitat: Humid tropical areas

Uses: Ornamental, Fodder, Medicinal, Antimicrobial

IUCN Threat listing: Not Assessed (NA)

Description: Evergreen shrub to small tree, up to 5 m high and profusely branched; Leaves alternate, bi-pinnate, pinnae 2, petioles long, slightly swollen at the base; Flowers red, blooming in puffs, with red silky stamens; Fruit brown, dry and hard pod.

Flowering and Fruiting: April-December

67. Scientific Name: *Callistemon citrinus* (Curtis) Skeels

Synonym: *Melaleuca citrina* (Curtis) Dum. Cours.

Family: Myrtaceae

Common Name: Lemon Bottle Brush

Vernacular Name: Cheel

Habitat: Open tropical areas

Uses: Hedge Plant, Ornamental

IUCN Threat listing: Not Assessed (NA)

Description: A shrub to small tree, 1-3 m high; Leaves 3-7 cm, veins clearly visible on both sides; Flowers in 6-10 cm long spikes, stamens red, purplish-red, or lilac, anthers dark-colored; Capsules in clusters along the stems, woody, cup-shaped.

Flowering and Fruiting: March-November

68. Scientific Name: *Calotropis procera* (Ait.) Ait. *f.*

Synonym: *Asclepias procera* Ait.

Family: Apocyanaceae

Common Name: Sodom Apple

Vernacular Name: Aakwan

Habitat: Open tropical grasslands

Uses: Medicinal

IUCN Threat listing: Not Assessed (NA)

Description: Middle-sized, erect or decumbent, milky shrub; Stem branched, with pale bark; Leaves thick, sub-sessile, ovate to obovate, acute, sub-amplexicaul; Flowers in terminal and axillary umbellate to corymbose cymes, purplish-red; Follicles recurved, smooth.

Flowering and Fruiting: Throughout the Year

69. Scientific Name: *Canavalia gladiata* (Jacq.)DC.

Synonym: *Dolichos gladiatus* Jacq.

Family: Leguminosae

Common Name: Sword bean

Vernacular Name: Ban Semi

Habitat: Humid tropical areas

Uses: Edible, Medicinal, Fodder

IUCN Threat listing: Not Assessed (NA)

Description: A perennial creeper, up to length of 10 m, stems brown, woody with age; Leaves alternate, large, trifoliolate, leaflets oval 7.5-20 cm long; Flowers papillonaceous, inverted, white to pink in colour; Fruits long, 20-60 cm, straight, rough-surfaced and slightly compressed, dehiscent pods, seeds 8-16.

Flowering and Fruiting: April-October

70. Scientific Name: *Canna indica* L.

Synonym: *Canna coccinea* Mill.

Family: Cannaceae

Common Name: Indian shot

Vernacular Name: Bajranti

Habitat: Open tropical areas

Uses: Ornamental, Medicinal, Fodder, Dye

IUCN Threat listing: Not Assessed (NA)

Description: A perennial, erect herb, up to 3.5 m high, rhizomes horizontally branching; Leaves spirally arranged, sheaths large, open, blade narrowly ovate to narrowly elliptical; Inflorescence terminal, racemose, usually simple, sometimes branched, flowers single or paired, irregular, bisexual flowers, pale red to yellow, androecium petaloid, forming the showy part of flower, of 5 staminodes, in 2 whorls; Fruit a loculicidal, dehiscent, ovoid capsule, softly spiny outside; seeds numerous, globose.

Flowering and Fruiting: August-October

71. Scientific Name: *Cardamine hirsuta* L.

Synonym: *Cardamine umbrosa* Andrz. *ex* DC.

Family: Brassicaceae

Common Name: Hairy Bitter Cress

Vernacular Name: Kadu Jadi

Habitat: Humid tropical areas

Uses: Edible

IUCN Threat listing: Not Assessed (NA)

Description: Annual, sub-erect, hirsute herb; Stems thin, slender, 8-30 cm; Leaves multilobed, basal 7-11, upper 5-9-lobed, ovate-lanceolate, sinuate-dentate to entire; Flowers small, in racemes, cream white; Fruit siliqua, ±1 cm long, erect, obtuse.

Flowering and Fruiting: May-August

72. Scientific Name: *Careya arborea* Roxb.

Synonym: *Barringtonia arborea* F. Muell.

Family: Lecythidaceae

Common Name: Wild Guava

Vernacular Name: Kumbhi

Habitat: Humid tropical areas

Uses: Edible, Medicinal

IUCN Threat listing: Not Assessed (NA)

Description: Small to medium-sized, deciduous tree, with spreading branches; Leaves up to 30 cm long, broadly obovate or obovate-oblong, rounded or shortly acuminate; Flowers large, white, ill-smelling, sessile, in thick, swollen, hard, few-flowered terminal spikes; Fruit 6-8 cm across, globose.

Flowering and Fruiting: February-July

73. Scientific Name: *Carissa spinarum* L.

Synonym: *Jasminonerium densiflorum* (Baker) Kuntze

Family: Apocyanaceae

Common Name: Carrisse

Vernacular Name: Karonda

Habitat: Open humid tropical areas

Uses: Edible, Pickle, Medicinal, Antimicrobial

IUCN Threat listing: Not Assessed (NA)

Description: Small, evergreen, armed shrub; Stem with light-grey bark, branchlets green, spined at base; Leaves opposite, ovate, acute, mucronate, dark green, shining and coriaceous; Flowers white, faintly-scented, in few-flowered cymes at the ends of branches; Berries sub-globose or elliptic, shining, dark purple when ripe.

Flowering and Fruiting: March-July

74. Scientific Name: *Caryota urens* L.

Family: Arecaceae

Common Name: Fish tail palm

Vernacular Name: Sulphi

Habitat: Humid tropical areas

Uses: Avenue tree, Beer making, Edible

IUCN Threat listing: Least Concerned (LC)

Description: An erect, tall palm, 16-20 m high, trunk smooth with prominent annular leaf-scars; Leaves bi-pinnate, 4-6 m long, pinnae 5-7 pairs, leaflets broadly cuneate, fan-shaped; Flowers many, in triads with female flower in the middle; Fruit globose, reddish purple; Seeds plano-convex, sub-reniform.

Flowering and Fruiting: January-April

75. Scientific Name: *Cascabela thevetia* (L.) H. Lippold

Synonym: *Thevetia peruviana* (Pers.) K. Schum.

Family: Apocyanaceae

Common Name: Yellow Oleander

Vernacular Name: Pila Kaner

Habitat: Humid tropical areas

Uses: Ornamental, Medicinal, Pest repellent

IUCN Threat listing: Not Assessed (NA)

Description: Evergreen shrub to small tree, up to 7 m high, stem erect, covered with light grayish bark; Leaves simple, alternate, sessile; Flowers terminal, funnel shaped, yellow in colour; Fruit is woody and rhomboid drupe.

Flowering and Fruiting: Throughout the Year

76. Scientific Name: *Catharanthus pusillus* (Murr.) G. Don

Synonym: *Vinca pusilla* Murray

Family: Apocyanaceae

Common Name: Tiny periwinkle

Vernacular Name: Sangkhi

Habitat: Open tropical areas

Uses: Medicinal

IUCN Threat listing: Not Assessed (NA)

Description: Small, erect, annual herb, branches quadrangular, spreading from the base; Leaves oppositely arranged, lance-shaped, 3-8 cm long, with rough margins; Flowers small, white, appear singly or in pairs, in the upper leaf axils; Seed pod is slender, 3-5 cm long, straight, pointed.

Flowering and Fruiting: May-July

77. Scientific Name: *Catunaregam spinosa* (Thunb.) Tirveng.

Synonym: *Randia spinosa* (Thunb.) Poir.

Family: Rubiaceae

Common Name: Mountian pomegranate

Vernacular Name: Mainhar

Habitat: Humid tropical areas

Uses: Edible, Fish Poison, Medicinal

IUCN Threat listing: Not Assessed (NA)

Description: Large deciduous, spiny shrub/ small tree; Stem with grey bark and horizontal spines; Leaves broader towards the apex, shortly petioled, glabrous or pubescent; Flowers yellowish or greenish-white, hairy, fragrant, solitary or 2-3 at the apices; Fruit a soft, fleshy berry, globose or ovoid.

Flowering and Fruiting: March-June

78. Scientific Name: *Celosia argentea* L.

Synonym: *Celosia cristata* L.

Family: Amaranthaceae

Common Name: Silver Cock's comb

Vernacular Name: Kurdu

Habitat: Open tropical areas

Uses: Ornamental, Edible

IUCN Threat listing: Not Assessed (NA)

Description: Erect, annual, herbs; Stems ridged; Leaves petioled, lanceolate or elliptic, glabrous, entire, apex acute, base acute-acuminate; Flowers borne in cylindric terminal spikes, pale-pink when young, becoming white and glistening. Capsules ellipsoid.

Flowering and Fruiting: October-December

79. Scientific Name: *Celosia spicata* (Thouars) Spreng.

Synonym: *Lestibudesia spicata* Thouars

Family: Amaranthaceae

Common Name: Flamingo pink

Vernacular Name: Safed Murga

Habitat: Open humid tropical areas

Uses: Ornamental

IUCN Threat listing: Not Assessed (NA)

Description: Small, annual, erect herb with grooved stem; Leaves alternate, linear-lanceolate, acute, entire, base tapering into the petiole; Flowers pink at first, turn white at maturity, stamens 5, united at base forming a cup, ovary superior and single celled; Fruit capsule, dehisce transversely, seeds 4-8, flat.

Flowering and Fruiting: October-December

80. Scientific Name: *Cenchrus biflorus* Roxb.

Synonym: *Cenchrus annularis* Andersson

Family: Poaceae

Common Name: Indian Sandbur

Vernacular Name: Bhurat

Habitat: Open tropical grasslands

Uses: Fodder

IUCN Threat listing: Not Assessed (NA)

Description: Annual, erect or ascending, tufted grass; Culms glabrous; Leaves linear-lanceolate, acuminate, rough with hairs; Racemes solitary, cylindrical, green, turning to black on maturity; Spikelets in clusters of 2-3, sessile, ovate, acute, bristly.

Flowering and Fruiting: October-January

81. Scientific Name: *Cestrum nocturnum* L.

Synonym: *Cestrum multiflorum* Roem. & Schult.

Family: Solanaceae

Common Name: Night blooming Jasmine

Vernacular Name: Raat ki Rani

Habitat: Open tropical areas

Uses: Ornamental, Aromatic, Medicinal

IUCN Threat listing: Not Assessed (NA)

Description: Sub-scandent, evergreen shrub; Stem profusely branched, pale green; Leaves alternate, lanceolate, acuminate, pale yellowish-green, shining; Flowers greenish-yellow, in axillary or terminal, leafy panicles, fragrant at night. Berries globose, turning white on maturity.

Flowering and Fruiting: July-November

82. Scientific Name: *Chamaecrista mimosoides* (L.)Greene

Synonym: *Cassia mimosoides* L.

Family: Leguminosae

Common Name: Feather leaved Cassia

Vernacular Name: *Patwa Ghas*

Habitat: Humid tropical areas

Uses: Fodder, Medicinal

IUCN Threat listing: Not Assessed (NA)

Description: Low, diffuse, glabrous or pubescent perennial herb; Stem with usually simple, sometimes spreading branches; Leaves 8-10 cm, leaflets 40-60 pairs on a puberulous rachis, linear, glabrous; Flowers solitary, sometimes 2-3 together, axillary; Pods flat, 5 cm long, linear, 15-20-seeded.

Flowering and Fruiting: July-December

83. Scientific Name: *Chionachne gigantea* (J.Koenig) Veldkamp

Synonym: *Coix gigantea* J. Jacq., *nom. illeg.*

Family: Poaceae

Common Name: Job's tear

Vernacular Name: Garmotika

Habitat: Humid tropical areas

Uses: Edible, Ornamental

IUCN Threat listing: Least Concerned (LC)

Description: A perennial, erect grass, up to 2 m high; Leaf-sheaths hispid, with tubercle-based hairs, ligule ciliate-membraneous, blades 25-60 cm long; Spikes monoecious, male and female spikelets in same inflorescence, glumes dissimilar, larger than florets, shiny, basal sterile florets with or without palea and oblong lemma; Caryopsis with adherent pericarp, 5-7 mm long, hilum punctiform.

Flowering and Fruiting: July-October

84. Scientific Name: *Chloris radiata* (L.) Sw.

Synonym: *Agrostis radiata* L.

Family: Poaceae

Common Name: Radiate fingergrass

Vernacular Name: Ungli Ghans

Habitat: Open tropical grasslands

Uses: Fodder

IUCN Threat listing: Not Assessed (NA)

Description: A gregarious, erect annual herb, stems erect or ascending, stems geniculate, decumbent, or lax, sometimes rooting at nodes; Leaves mostly cauline, sheathing at base, leaf sheath mostly open, or loose, smooth, glabrous; Inflorescence terminal, solitary, with 1 spike, flowers bisexual, spikelets sessile or sub-sessile, spikelets laterally compressed, spikelet less than 3 mm wide; Fruit caryopsis, ellipsoid, longitudinally grooved, hilum long-linear.

Flowering and Fruiting: March-September

85. Scientific Name: *Chrozophora rottleri* (Geiseler) A. Juss. *ex* Spreng.

Synonym: *Croton rottleri* Geiseler

Family: Euphorbiaceae

Common Name: Rottler's Chrozophora

Vernacular Name: Shahdevi

Habitat: Open tropical areas

Uses: Dye, Medicinal

IUCN Threat listing: Not Assessed (NA)

Description: Annual herbs, prostrate or ascending, main stem up to 50 cm long; Leaves alternate, 2-5 x 1-4 cm, rounded or obtuse at apex, base rounded or sub-truncate, entire or shallowly crenate-sinuate; Male flowers with smaller pedicels, female flowers with long pedicels, extending further more in fruit; Fruits 4 x 7 mm, rounded, 3-lobed, stellate-pubescent, seeds globose-ovoid, grey.

Flowering and Fruiting: February-May

86. Scientific Name: *Chrysopogon aciculatus* (Retz.) Trin.

Synonym: *Andropogon aciculatus* Retz.

Family: Poaceae

Common Name: Golden false beardgrass

Vernacular Name: Surwala

Habitat: Open tropical grasslands

Uses: Fodder, Medicinal

IUCN Threat listing: Not Assessed (NA)

Description: A rhizomatous or stoloniferous, spreading perennial, culms often erect, 15-25 cm tall, geniculate, branching, rooting at the nodes; Leaves 3-15 cm long, linear, hairy at mouth only; Flowering branch with group of 3 spikelets at its end; Caryopsis oblong, 2 mm long.

Flowering and Fruiting: June-October

87. Scientific Name: *Cinnamomum camphora* (L.) J. Presl

Synonym: *Laurus camphora* L.

Family: Lauraceae

Common Name: Camphor tree

Vernacular Name: Kapur

Habitat: Humid tropical areas

Uses: Aromatic, Medicinal, Cosmetics

IUCN Threat listing: Not Assessed (NA)

Description: A large, evergreen tree, with a stout, uniformly cylindrical bole and dense, symmetrical crown; Leaves simple, petiolate, alternate, ovate-elliptic, base and apex acuminate, light green, smooth and shining above, whitish or glaucous below; Flowers bisexual, membranous, linear, soft-pubescent; Fruit globular, 7-8 mm wide, purplish-black, maturing to a berry-like, one seeded drupe.

Flowering and Fruiting: April-November

88. Scientific Name: *Cleome monophylla* L.

Synonym: *Sieruela viscosa* Rafin.

Family: Cleomaceae

Common Name: Spindle pod

Vernacular Name: Hurhura

Habitat: Humid tropical areas

Uses: Edible, Medicinal

IUCN Threat listing: Not Assessed (NA)

Description: Erect, annual herb, up to 40 cm, stem branched, angled, and striate; Leaves simple, ovate-oblong, base truncate, margins ciliate; Racemes terminal, lax, up to 15 cm long, flower pedicels glandular pubescent, sepals linear, acuminate, petals pink or light purple; Capsule 7-10 cm long, linear, sub-cylindric, beaked, strongly ribbed, glandular, hairy.

Flowering and Fruiting: June-November

89. Scientific Name: *Clerodendrum indicum* (L.) Kuntze

Synonym: *Clerodendrum longicolle* G.Mey.

Family: Lamiaceae

Common Name: Tubeflower

Vernacular Name: Bharangi

Habitat: Humid tropical areas

Uses: Ornamental, Medicinal

IUCN Threat listing: Not Assessed (NA)

Description: Perennial shrub; Stem fluted, hollow; Leaves sub-sessile, in whorls of 3-5, lanceolate, glabrous, shining, apex and base acute; Flowers in rigid terminal panicles, white, fading to yellow; Drupes dark blue, 1-4-seeded, seated on accrescent red calyx.

Flowering and Fruiting: January-May

90. Scientific Name: *Clitoria ternatea* L.

Synonym: *Lathyrus spectabilis* Forssk.

Family: Leguminosae

Common Name: Butterfly pea

Vernacular Name: Aparajita

Habitat: Open tropical areas

Uses: Edible, Fodder, Medicinal

IUCN Threat listing: Not Assessed (NA)

Description: A perennial, herbaceous vine or creeper, doing well in moist, neutral soil; Leaves pinnate, with 5-7 elliptical, 3-5 cm long leaflets; Flowers solitary or paired, deep blue or pure white, about 4 cm broad; Pods flat, linear, sparsely pubescent, dehiscing violently at maturity, seeds 8-10 dark, shiny.

Flowering and Fruiting: July-September

91. Scientific Name: *Cocos nucifera* L.

Synonym: *Diplothemium henryanum* F.Br.

Family: Arecaceae

Common Name: Coconut

Vernacular Name: Nariyal

Habitat: Humid tropical areas

Uses: Edible, Mats, Broom, Ropes, Medicinal

IUCN Threat listing: Not Assessed (NA)

Description: An erect, unbranched evergreen, beautiful tree; Leaves pinnate, feather shaped, 4-7 m long and 1-1.5 m wide at the broadest part; Inflorescence consists of female and male axillary flowers, small and light yellow, in clusters that emerge from canoe-shaped sheaths among the leaves; Fruit a drupe, roughly ovoid, up to 30 cm long and 20 cm wide, composed of a thick, fibrous husk surrounding a somewhat spherical nut with a hard, brittle, hairy shell.

Flowering and Fruiting: Throughout the Year

92. Scientific Name: *Codiaeum variegatum* (L.) Rumph. *ex* A. Juss.

Synonym: *Croton variegatus* L.

Family: Euphorbiaceae

Common Name: Garden Croton

Vernacular Name: Kroton

Habitat: Humid tropical areas

Uses: Ornamental, Medicinal

IUCN Threat listing: Not Assessed (NA)

Description: Ornamental shrub, up to 2 m tall, stems white-brown, initially green; Leaves alternate, lanceolate, symmetrical, margin entire, mid-veins yellow, apex acute, base acute, green-yellow variegation on lamina; Inflorescence axillary raceme or panicle, bracts reduced, flowers unisexual.

Flowering and Fruiting: September-November

93. Scientific Name: *Colebrookea oppositifolia* Sm.

Synonym: *Elsholtzia oppositifolia* (Sm.) Poir.

Family: Lamiaceae

Common Name: Indian Squirrel tail

Vernacular Name: Bhirmoli

Habitat: Open tropical areas

Uses: Fodder, Medicinal

IUCN Threat listing: Not Assessed (NA)

Description: An erect, hoary, much branched shrub; Stem up to 3 m high, branches nearly quadrangular; Leaves opposite, hairy on both surfaces, oblong-elliptic, acuminate, crenate-serrate; Flowers greenish-white, in congested clusters at the end of branches; Nutlets hairy, obovoid.

Flowering and Fruiting: January-May

94. Scientific Name: *Colocasia esculenta* (L.) Schott

Synonym: *Alocasia esculenta*

Family: Araceae

Common Name: Taro

Vernacular Name: *Kochai*

Habitat: Humid tropical areas

Uses: Edible, Medicinal

IUCN Threat listing: Least Concerned (LC)

Description: Fast growing, perennial herb, up to 1.5 m, with a massive fleshy corm at the base, and lateral, thick, edible runners; Leaves arranged in a loose rosette, blades pointing downward, cordate or lanceolate, sub-coriaceous, green above, glaucous below; Spadix yellow, much shorter than the spathe, sterile flower zone and the distal appendage shorter than the fertile zone; Fruit many-seeded berry, densely packed and forming a fruiting head.

Flowering and Fruiting: August-October

95. Scientific Name: *Combretum indicum* (L.) C. C. H. Jongkind

Synonym: *Quisqualis indica* L.

Family: Combretaceae

Common Name: Rangoon creeper

Vernacular Name: Madhu Malti

Habitat: Humid tropical areas

Uses: Ornamental, Aromatic, Medicinal

IUCN Threat listing: Not Assessed (NA)

Description: A deciduous, creeping plant, up to 6 m long, stem much branched from the base, branches cylindrical, somewhat compressed at nodes; Leaves opposite or sub-opposite, elliptical, oblong, or lanceolate; Flowers sessile or sub-sessile, in terminal spikes or racemes, on short lateral branches, petals 5, oblong or oblanceolate, pink, turning red when mature; Fruit elliptical in outline, angular, with 5 narrow wings.

Flowering and Fruiting: March-August

96. Scientific Name: *Commelina diffusa* Burm. *f.*

Synonym: *Commelina agraria* Kunth

Family: Commelinaceae

Common Name: Climbing dayflower

Vernacular Name: Kana

Habitat: Humid tropical areas

Uses: Fodder, Medicinal

IUCN Threat listing: Least Concerned (LC)

Description: Smooth, sparsely hairy, annual or perennial herb, stem creeping; Leaves lanceolate to broadly lanceolate, 4-6 cm long, gradually acute to acuminate; Flowers open only in the morning, blue, fertile stamens 3, sterile 2 (rarely 3); Fruit a 3-celled capsule, 4-5 mm long, seeds 5.

Flowering and Fruiting: July-October

97. Scientific Name: *Convolvulus equitans* Benth.

Synonym: *Convolvulus hermanniae* Choisy

Family: Convolvulaceae

Common Name: Texas Bindweed

Habitat: Open tropical grasslands

Uses: Medicinal

IUCN Threat listing: Not Assessed (NA)

Description: Annual, creeping herb, up to 6 feet long; Leaves variable, narrowly triangular or nearly linear, may have two symmetric groups of angular lobes at the base; Flowers white to pale pink or bluish purple, with red or pale yellow centre; Fruits are orange/brown spheres, nearly 1 cm wide.

Flowering and Fruiting: April-October

98. Scientific Name: *Convolvulus prostratus* Forsk.

Synonym: *Convolvulus microphyllus* Sieb. *ex* Spreng.

Family: Convolvulaceae

Common Name: Prostrate Bindweed

Vernacular Name: Shankhpushpi

Habitat: Open tropical grasslands

Uses: Medicinal, Oil

IUCN Threat listing: Not Assessed (NA)

Description: A diffuse, hairy annual or perennial herb; Stem and branches slender, prostrate or sub-erect; Leaves short-petioled, linear-oblanceolate, apex mucronate, sparsely hairy on both surfaces; Flowers solitary or in pairs, sessile or sub-sessile in the leaf axils, pale pinkish-white; Capsules globose, hairy.

Flowering and Fruiting: November-January

99. Scientific Name: *Corchorus fascicularis* Lam.

Synonym: *Corchorus brachycarpus* Guill. & Perr.

Family: Malvaceae

Common Name: Wild Jute

Vernacular Name: Sanai

Habitat: Open tropical areas

Uses: Edible, Fibre, Medicinal

IUCN Threat listing: Not Assessed (NA)

Description: An annual, sub-erect herb, often 50-60 cm high, stems woody, bark scaly; Leaves 3-costate, elliptic-oblong or lanceolate, glabrous, serrate; Flowers in fascicles, yellow, 5-6 mm across, bracts minute, sepals linear-oblong, apiculate; Capsules 2-5(-8) per group, sessile, 1-1.5 cm long, beaked, somewhat triangular, pubescent.

Flowering and Fruiting: August-February

100. Scientific Name: *Cordia myxa* L.

Synonym: *Cordia obliqua* Willd.

Family: Boraginaceae

Common Name: Assyrian Plum

Vernacular Name: Lasora

Habitat: Humid tropical areas

Uses: Edible, Fuel wood, Medicinal, Fibre

IUCN Threat listing: Not Assessed (NA)

Description: Medium-sized, evergreen tree, often reaching a height up to 6 m; Leaves petiolate, blade ovate-lanceolate, broadly ovate, rarely suborbicular; Flowers vary in number per cluster, creamy white, stamens long and exserted; Fruit ovoid drupe, seeds contained in pulp.

Flowering and Fruiting: May-August

101. Scientific Name: *Corynandra viscosa* L.

Synonym: *Cleome viscosa* L.

Family: Cleomaceae

Common Name: Asian Spider flower

Vernacular Name: Hulhul

Habitat: Humid tropical areas

Uses: Wound healing, Medicinal

IUCN Threat listing: Not Assessed (NA)

Description: Erect, profusely branching, viscous, annual herb, covered with stalked glandular hairs; Leaves 3-5(-7)-foliolate, leaflets sub-sessile, elliptic-oblong or obovate to spathulate, base cuneate, margins ciliate; Flowers solitary on long axillary stalks, further elongating during fruiting, sepals 4, petals 4; Capsules linear-oblong, terete, striate, densely glandular hairy.

Flowering and Fruiting: July-December

102. Scientific Name: *Cosmos bipinnatus* Cav.

Synonym: *Dahlia repens* Hartw. *ex* Benth.

Family: Asteraceae

Common Name: Garden Cosmos

Habitat: Open tropical areas

Uses: Ornamental, Medicinal

IUCN Threat listing: Not Assessed (NA)

Description: An erect, annual, ornamental herb, up to 3 m, stems slender, angled, nodes more or less cupulate; Leaves opposite, petiolate, bipinnate; Flower heads white, pink or red, corolla lobes truncate or dentate; Achenes glabrescent, linear, ribbed, with an abrupt beak.

Flowering and Fruiting: June-December

103. Scientific Name: *Cosmos caudatus* Kunth

Synonym: *Bidens caudata* (Kunth) Sch.Bip.

Family: Asteraceae

Common Name: Wild Cosmos

Habitat: Open tropical areas

Uses: Ornamental, Medicinal

IUCN Threat listing: Not Assessed (NA)

Description: Annual, erect herb, with a weak taproot, up to 2 m high, stems branched, glabrous or sparsely hispid; Leaves cauline, alternate, deeply lobed, 10-20 cm long; Capitulae 5-15 mm across, involucral bracts erect, oblong-lanceolate, ray florets rose-pink to purple, disc florets 5-6 mm long; Cypselae light brown, glabrous or scabridulous proximally, setose distally.

Flowering and Fruiting: October-February

104. Scientific Name: *Cosmos sulphureus* Cav.

Synonym: *Bidens artemisiifolia* (Jacq.) Kuntze

Family: Asteraceae

Common Name: Sulphur Cosmos

Habitat: Open tropical areas

Uses: Ornamental, Medicinal

IUCN Threat listing: Not Assessed (NA)

Description: An annual, erect herb, taproot strong, up to 2 m tall, branches glabrous or sparsely pilose to hispid; Leaves cauline, alternate, deeply lobed 5-25 cm long; Capitulae 6-10 mm across, involucral bracts erect, oblong-lanceolate, ray florets golden yellow to red-orange; Cypselae light brown, flattened, hispidulous, rarely glabrous.

Flowering and Fruiting: September-December

105. Scientific Name: *Crassocephalum crepidioides* (Benth.) S.Moore

Synonym: *Gynura crepidioides* Benth.

Family: Asteraceae

Common Name: Red-flower Ragleaf

Habitat: Open tropical grasslands

Uses: Edible, Medicinal

IUCN Threat listing: Not Assessed (NA)

Description: Erect, sparingly branched, aromatic, annual herb, 40-100 cm tall, stem stout, soft, ribbed, hairy above, glabrescent below; Leaves helically arranged, elliptic, oblong or obovate-elliptic, acute or acuminate, pinnately lobed or pinnatifid, irregularly serrate; Flowers equal, bisexual, corolla yellow throughout, 9-11 mm long, tubular, tube long, very slender, funnel-shaped; Achenes cylindric-linear, ribbed, dark-brown with paler base and apex, thinly pubescent.

Flowering and Fruiting: August-December

106. Scientific Name: *Crateva magna* (Lour.) DC

Synonym: *Crateva nurvala* Buch.-Ham.

Family: Capparaceae

Common Name: Varuna

Vernacular Name: Varun

Habitat: Humid tropical areas

Uses: Medicinal, Sacred

IUCN Threat listing: Not Assessed (NA)

Description: A deciduous tree; Stem with grey, fairly smooth bark; Leaves palmate, leaflets 3, obovate- oblong or lanceolate, acuminate, entire, glabrous; Flowers white, pale-yellow or light rose, in many-flowered, lax, terminal corymbs; Fruit a many seeded ovoid berry, seeds embedded in a yellow pulp.

Flowering and Fruiting: March-April

107. Scientific Name: *Crinum asiaticum* (R.Br.) Fosberg & Sachet

Synonym: *Crinum pedunculatum* R.Br.

Family: Amaryllidaceae

Common Name: Spider Lily

Vernacular Name: Sudarshan

Habitat: Humid tropical areas

Uses: Medicinal

IUCN Threat listing: Not Assessed (NA)

Description: Perennial, bulbous herbs; Leaves clustered at bulb apex, oblong, flat, coriaceous, glabrous, narrow at base, margin entire, apex gradually tapering; Flowers up to 16.5 cm long, perianth salver-shaped, lobes white, oblong-linear; Ovary 3-celled, ovule(s) 1 or 2.

Flowering and Fruiting: August-December

108. Scientific Name: *Crotalaria calycina* Schrank

Synonym: *Crotalaria roxburgiana* DC.

Family: Leguminosae

Common Name: Hairy Rattlepod

Habitat: Open tropical grasslands

Uses: Medicinal

IUCN Threat listing: Not Assessed (NA)

Description: An annual herb, 30-60 cm; Leaves mostly 5-8 cm, linear ones up to 15 cm long, acute, glabrous, obscurely silky beneath; Flowers 2-12, in lax terminal racemes, sepals and bracts hairy, corolla pale yellow, shorter than the calyx; Pod linear-oblong, sessile, glabrous, 1.2-2 cm long.

Flowering and Fruiting: August-March

109. Scientific Name: *Crotalaria juncea* L.

Synonym: *Crotalaria benghalensis* Lam.

Family: Leguminosae

Common Name: Sunn Hemp

Vernacular Name: Jhunjhuna

Habitat: Open tropical grasslands

Uses: Fiber, Fodder, Manure, Medicinal

IUCN Threat listing: Not Assessed (NA)

Description: An erect, annual herb, 60-240 cm, branches numerous, ascending, striate, silky-pubescent; Leaves 2.5-10 cm long, linear or oblong, appressed silky; Flowers large, yellow, in erect terminal and lateral 12-20 flowered racemes; Pods 2.5- 3 cm long, sessile, inflated, clothed with short fulvous silky hairs.

Flowering and Fruiting: July-September

110. Scientific Name: *Crotalaria medicaginea* Lam.

Synonym: *Crotalaria foliosa* Willd.

Family: Leguminosae

Common Name: Blue Rattlepod

Habitat: Open tropical grasslands

Uses: Medicinal

IUCN Threat listing: Not Assessed (NA)

Description: An erect, perennial, pubescent herb or undershrub herb; Stem with several, ascending branches; Leaves trifoliate, leaflets variable, sub-sessile, oblanceolate, pubescent beneath, apex mucronate; Flowers yellow, in many-flowered, erect, often leaf-opposed racemes; Pods obliquely sub-globose, 2-seeded.

Flowering and Fruiting: August-February

111. Scientific Name: *Crotalaria pallida* Aiton

Synonym: *Crotalaria mucronata* Desv

Family: Leguminosae

Common Name: Smooth Rattlebox

Vernacular Name: San

Habitat: Open tropical grasslands

Uses: Medicinal

IUCN Threat listing: Not Assessed (NA)

Description: Erect, robust herb, up to 2 m high; Leaves 3-foliolate with elliptic to obovate leaflets, appressed puberulous below; Flowers in dense, erect, spiky clusters, pale lemon-yellow; Pods inflated, pendent, up to 5 cm.

Flowering and Fruiting: September-January

112. Scientific Name: *Crotalaria retusa* L.

Synonym: *Crotalaria hostmannii* Steud.

Family: Leguminosae

Common Name: Rattleweed

Vernacular Name: Ghunghuniya

Habitat: Open tropical grasslands

Uses: Fiber, Manure, Medicinal

IUCN Threat listing: Not Assessed (NA)

Description: An annual, erect herb, up to 130 cm; Leaves simple, oblanceolate to spatulate, glabrous above, densely seriaceous pubescent below; Racemes few-flowered, bracts narrowly triangular, calyx broadly bell-shaped, petals bright yellow, the standard ovate, clawed, with reddish veins; Pods dark brown to black at maturity, oblong, nearly cylindrical, glabrous.

Flowering and Fruiting: September-January

113. Scientific Name: *Crotalaria spectabilis* Roth

Synonym: *Crotalaria leschenaultii* DC.

Family: Leguminosae

Common Name: Showy Rattlebox

Vernacular Name: Sanni

Habitat: Open tropical grasslands

Uses: Edible, Manure, Fibre, Medicinal

IUCN Threat listing: Not Assessed (NA)

Description: An erect, biennial undershrub, up to 1.2 m tall with glabrous, furrowed branches; Leaves 5-15 cm, oblong-lanceolate, apex acute or subacute, mucronate, base cuneate, glabrous above, fine-silky beneath; Flowers borne in elongated 16-50 flowered terminal racemes, yellow with a purple tinge; Fruit (pod) stalked, linear-oblong, 2.5-5 cm long, glabrous, seeds 20-30.

Flowering and Fruiting: November-January

114. Scientific Name: *Cucumis callosus* (Rottb.) Cogn.

Synonym: *Cucumis melo* Blanco

Family: Cucurbitaceae

Common Name: Melon

Vernacular Name: Pihiti

Habitat: Open tropical grasslands

Uses: Edible, Medicinal

IUCN Threat listing: Not Assessed (NA)

Description: A perennial climber, 1.5-2.0 m long, stems slender, angular, rough with short, rigid hairs; Leaves alternate, 3-6 cm long, suborbicular, base cordate, palmately 5-7 lobed, lobes round or ovate-oblong; Male flowers often solitary, calyx-tube narrow campanulate, corolla yellow, staminal filaments short, female flowers in small groups, ovary densely hairy, stigmas converging; Fruits obovoid, oblong, white.

Flowering and Fruiting: July-September

115. Scientific Name: *Cullen corylifolium* (L.) Medik

Synonym: *Psoralea corylifolia* L.

Family: Leguminosae

Common Name: Scrufy Pea

Vernacular Name: Bakuchi

Habitat: Open tropical grasslands

Uses: Medicinal

IUCN Threat listing: Not Assessed (NA)

Description: A sub-erect, annual herb, up to 50-90 cm; Leaves simple, alternate, petiolate, ovate, dentate; Flowers in short, condensed, axillary spikes, pale-purple; Fruits small pods, one-seeded.

Flowering and Fruiting: August-December

116. Scientific Name: *Cuphea carthagenensis* (Jacq.) J.F. Macbr.

Synonym: *Cuphea balsamona* (Vand.) Cham. & Schltdl.

Family: Lyhtraceae

Common Name: Colombian Waxweed

Habitat: Humid tropical areas

Uses: Ornamental, Medicinal

IUCN Threat listing: Not Assessed (NA)

Description: An annual herb of moist habitats, stem viscid-pilose, with intermixed glandular and non-glandular hairs; Leaves opposite, sub-sessile to shortly petioled, elliptic, ovate, rarely obovate, with acute apex; Flowers solitary, axillary, floral tube with glandular hairs, green, calyx lobes unequal, petals 6, linear-elliptic, pale purple, stamens longer than the floral tube.

Flowering and Fruiting: June-November

117. Scientific Name: *Cuphea hyssopifolia* Kunth.

Synonym: *Parsonsia hyssopifolia* (Kunth.) Standl.

Family: Lyhtraceae

Common Name: Mexican Heather

Habitat: Open tropical areas

Uses: Ornamental, Medicinal

IUCN Threat listing: Not Assessed (NA)

Description: An erect to spreading, low-branching, compact, free-flowering herb; Leaves simple, sub-sessile, opposite, pinnately-veined, oblong, sometimes elliptic, glossy; Flowers solitary, axillary, trumpet-shaped, sepals crinkled, rarely absent, petals 6, purple, floral tube pale green, glabrous or slightly pubescent; Fruits leathery, dehiscent, oblong-ovoid, 1-celled.

Flowering and Fruiting: Throughout the Year

118. Scientific Name: *Curculigo orchioides* Gaertn.

Synonym: *Hypoxis orchioides* (Gaertn.) Kurz

Family: Hypoxidaceae

Common Name: Golden eye grass

Vernacular Name: Kali Musli

Habitat: Humid tropical areas

Uses: Medicinal

IUCN Threat listing: Not Assessed (NA)

Description: Small, annual, ephemeral herbs, rhizome oblong, perennial; Leaves 10-15 x 2 cm, lanceolate, plicate, base sheathing, pilose, subsessile; Flowers on long pedicel from the base, 1 or rarely few, perianth yellow, tube narrow, stamens 6, ovary 3-celled; Fruit baccate, seeds sub-globose.

Flowering and Fruiting: June-December

119. Scientific Name: *Curcuma caesia* Roxb.

Synonym: *Curcuma kuchoor* Royle

Family: Zingiberaceae

Common Name: Black Turmeric

Vernacular Name: Kali Haldi

Habitat: Humid tropical areas

Uses: Spice, Cosmetic, Medicinal

IUCN Threat listing: Not Assessed (NA)

Description: A leafy, rhizomatous herb, rhizome blue inside; Leaves 30-60 cm long, broadly lanceolate-oblong, with a broad purple cloud along the midrib; Flowers in compact, oblong spikes, appearing before leaves, pale yellow, reddish at the outer border.

Flowering and Fruiting: August-October

120. Scientific Name: *Cyanotis vaga* (Lour.) Schult. & Schult. *f.*

Synonym: *Tonningia vaga* (Lour.) Kuntze

Family: Commelinaceae

Common Name: Wandering dew grass

Habitat: Humid tropical areas

Uses: Fodder, Medicinal

IUCN Threat listing: Not Assessed (NA)

Description: Annual, prostrate, terrestrial herb, up to 70 cm long, stem rounded, solid, glabrous, succulent; Leaves simple, not divided or lobed, alternate, spiral, sessile, glabrous on both sides, margin entire, apex acute, base clasping, parallel-veined; Flowers bisexual, in axillary or terminal inflorescences, covered by a spathe, single or few, sessile, blue, petals 3; Fruit a capsule, opening with 3 valves.

Flowering and Fruiting: September-January

121. Scientific Name: *Cynodon dactylon* (L.) Pers.

Synonym: *Agrostis bermudiana* Tussac. *ex* Kunth

Family: Poaceae

Common Name: Bermuda Grass

Vernacular Name: Dubi

Habitat: Open tropical grasslands

Uses: Fodder, Medicinal

IUCN Threat listing: Not Assessed (NA)

Description: A perennial grass; Stem extensively creeping, prostrate, rooting at the nodes, mat forming; Leaves linear, acuminate; Inflorescence terminal, digitate, of 3-6 green or purplish spikes; Spikelets in two rows along the axis, laterally compressed.

Flowering and Fruiting: March-October

122. Scientific Name: *Cynoglossum wallichii* G. Don

Synonym: *Cynoglossum calycinum* Wall.

Family: Boraginaceae

Common Name: Imertian Hound's tongue

Vernacular Name: Kamraj

Habitat: Open tropical areas

Uses: Medicinal

IUCN Threat listing: Not Assessed (NA)

Description: Robust, biennial herb, up to 75 cm; Cauline leaves petiolate, elliptic, middle ones largest, sessile and decurrent on stem by broad base, oblong elliptic; Inflorescence narrow when young, branches 7-12 flowered, finally erecto patent, petals black or dark purple, calyx velvety, enlarging in fruit, longer than nutlets, shortly strigose.

Flowering and Fruiting: May-August

123. Scientific Name: *Cyperus difformis* L.

Synonym: *Cyperus subrotundus* Llanos

Family: Cyperaceae

Common Name: Flat Sedge

Vernacular Name: Bada Naagar-motha

Habitat: Open tropical grasslands

Uses: Fodder, Medicinal

IUCN Threat listing: Least Concerned (LC)

Description: An aquatic perennial herb, height range 6-80 cm, stems smooth, triangular, slightly winged; Leaves smooth or slightly scabrid on the midrib, flat, linear, often two-thirds of the plant height, sometimes reduced to sheaths; Inflorescence dense, globose, umbellate heads, simple or compound, with 10-60 stellately spreading spikelets.

Flowering and Fruiting: July-November

124. Scientific Name: *Cyperus esculentus* L.

Synonym: *Chlorocyperus aureus* (K.Richt.) Palla *ex* Kneuck.

Family: Cyperaceae

Common Name: Nut Sedge

Habitat: Open tropical grasslands

Uses: Edible, Oil, Comsetic, Medicinal

IUCN Threat listing: Least Concerned (LC)

Description: Erect perennial herb, culms simple with triangular section, growing from perennial, tuber-bearing rhizomes; Leaves in 3 ranks, mostly basal; Inflorescence terminal umbellate, umbels subtended by unequal leafy bracts, spikelets yellowish-brown 1-3 cm, many flowered, flattened, stamens 3, style 3-cleft; Achenes 3-angled.

Flowering and Fruiting: June-October

125. Scientific Name: *Cyperus iria* L.

Synonym: *Chlorocyperus iria* (L.) Rikli

Family: Cyperaceae

Common Name: Rice Flatsedge

Habitat: Open tropical grasslands

Uses: Fodder, Medicinal

IUCN Threat listing: Least Concerned (LC)

Description: Annual, erect, glabrous herb; Culms triquetrous; Leaves linear-lanceolate, acuminate, sheaths brown, linear-lanceolate; Inflorescence a spike, spikelets oblong, obtuse; Nuts trigonous, black.

Flowering and Fruiting: August-November

126. Scientific Name: *Cyperus rotundus* L.

Synonym: *Chlorocyperus rotundus* (L.) Palla

Family: Cyperaceae

Common Name: Common Nut Sedge

Vernacular Name: Nagar Motha

Habitat: Open tropical grasslands

Uses: Edible, Medicinal, Mats

IUCN Threat listing: Least Concerned (LC)

Description: An erect herb, of moist and waterlogged areas, stems erect, up to 60 cm tall, 3-sided, smooth with swollen bases, rhizomes wiry, dark and persistent, giving many daughter shoots and tubers; Inflorescence terminal, open umbel subtended by several leafy bracts, spikelets 3-8, reddish-brown, flattened, each with up to 30 glumes; Fruit a 3-angled achene, 1.5 mm long, dark brown or black.

Flowering and Fruiting: March-August

127. Scientific Name: *Dactyloctenium aegyptium* (L.) Willd.

Synonym: *Chloris mucronata* Michx.

Family: Poaceae

Common Name: Crowfoot grass

Vernacular Name: Makra

Habitat: Open tropical grasslands

Uses: Edible, Fodder

IUCN Threat listing: Not Assessed (NA)

Description: Annual, sub-erect or ascending herb; Leaves linear-lanceolate, distichous, fine-pointed, hairy; Inflorescence of 4-6 cm long, radiate spikes of 4-5 spikelets; Spikelets densely crowded, ovate, mucronate, awned.

Flowering and Fruiting: August-February

128. Scientific Name: *Dalbergia sissoo* DC.

Synonym: *Amerimnon sissoo* (Roxb.)Kuntze

Family: Leguminosae

Common Name: Indian Rosewood

Vernacular Name: Sesam

Habitat: Open tropical areas

Uses: Fuel wood, Timber, Medicinal

IUCN Threat listing: Not Assessed (NA)

Description: Very large, much branched tree; Trunk with grey, longitudinally furrowed bark; Leaves usually trifoliate, leaflets 3-5, alternate, sub-orbicular, rhomboid, apex acuminate or acute, rachis zigzag; Flowers creamish-white, in short, axillary panicles; Pods strap-shaped, 1-3-seeded.

Flowering and Fruiting: March-June

129. Scientific Name: *Datura metel* L.

Synonym: *Stramonium datura* Noronha

Family: Solanaceae

Common Name: Angel's trumpet

Vernacular Name: Safed Dhatura

Habitat: Humid tropical areas

Uses: Medicinal

IUCN Threat listing: Not Assessed (NA)

Description: An erect perennial; Stem woody below, somewhat purple near the tips, branches zigzag; Leaves ovate-lanceolate, acuminate, sinuate, oblique at base and with unequal sides; Flowers pedicelled, white or purple; Capsules globose, covered with blunt spines or tubercles.

Flowering and Fruiting: March-December

130. Scientific Name: *Datura stramonium* L.

Synonym: *Datura praecox* Godr.

Family: Solanaceae

Common Name: Jimson weed

Vernacular Name: Safed Dhatura

Habitat: Open humid tropical areas

Uses: Medicinal, Pest repellent

IUCN Threat listing: Not Assessed (NA)

Description: An erect, glabrous, annual herb; Stem 60-80 cm high; Leaves stalked, ovate, irregularly lobed and serrate; Flowers white, solitary, axillary, corolla funnel-shaped; Capsules ovoid, covered with rigid, sharp prickles, surrounded at base by enlarged calyx lobes.

Flowering and Fruiting: July-January

131. Scientific Name: *Delonix regia* (Hook.) Raf.

Synonym: *Poinciana regia* Hook.

Family: Leguminosae

Common Name: Flamboyant

Vernacular Name: Gulmohar

Habitat: Open tropical areas

Uses: Avenue tree, Fuel wood, Timber, Medicinal

IUCN Threat listing: Least Concerned (LC)

Description: A tall tree with spreading crown, may reach up to 15 m; Leaves compound, bipinnate and feathery, up to 60 cm long, pinnae 11-18 pairs, petiole stout, leaflets 20-30 pairs on each pinna, oblong; Flowers borne in lax, terminal or axillary racemes at the end of branches, large, 10 cm across, bright red; Pods 5 cm broad, 30-60 cm long, ending in a beak when mature.

Flowering and Fruiting: April-July

132. Scientific Name: *Dendrocalamus hamiltonii* Nees & Arn. *ex* Munro

Synonym: *Sinocalamus hamiltonii* (Nees & Arn. *ex* Munro) T.Q. Nguyen

Family: Poaceae

Common Name: Tama bamboo

Vernacular Name: Baans

Habitat: Humid tropical areas

Uses: Edible, Fibre, Handicrafts

IUCN Threat listing: Not Assessed (NA)

Description: Deciduous, densely tufted bamboo, culms solid, 7-17 m, 5-10 cm in diameter, internodes white powdery; Spikelets densely pubescent, fertile florets 2-4, glumes 2-more, long mucronate, anthers yellow; Caryopsis brown, shining, ovoid to sub-globose, hairy above, beaked with persistent style base.

Flowering and Fruiting: Monocarpic

133. Scientific Name: *Dendrocalamus membranaceus* Munro

Synonym: *Bambusa membranacea* (Munro) Stapleton & N.H. Xia

Family: Poaceae

Common Name: Solid Bamboo

Vernacular Name: Katang Baans

Habitat: Humid tropical areas

Uses: Paper making, Fibre, Handicraft

IUCN Threat listing: Least Concerned (LC)

Description: A large, evergreen, clump-forming bamboo with erect stems; Leaves to 16 x 1.5 cm, ovate-lanceolate, apex long acuminate, spinescent, ligule dentate; Heads of spikelets globose, spikelets ovate, glumes ovate, lemmas 2, broadly ovate with a mucro, stamens 6, filaments free, anthers oblong.

Flowering and Fruiting: Monocarpic

134. Scientific Name: *Desmodium laxiflorum* DC.

Synonym: *Desmodium diffusum* DC.

Family: Leguminosae

Common Name: 3-flower Tick Trefoil

Vernacular Name: Kudaliya

Habitat: Open tropical areas

Uses: Medicinal

IUCN Threat listing: Not Assessed (NA)

Description: An erect, 30-60 cm high undershrub; Stem and young branches densely clothed with short hairs; Leaves trifoliate, leaflets ovate or elliptic, acute, clothed below with minute appressed hairs; Flowers in axillary or terminal drooping, branched racemes, whitish-yellow, tinged with blue; Pods oblong, flattened, covered with minute, hooked hairs.

Flowering and Fruiting: August-October

135. Scientific Name: *Dianthus woroschilovii* Barkalov & Prob.

Synonym: *Dianthus chinensis* subsp. *reflexus* Worosch.

Family: Caryophyllaceae

Common Name: Chinese Pink

Habitat: Open tropical areas

Uses: Ornamental, Medicinal

IUCN Threat listing: Not Assessed (NA)

Description: Perennial herbs, about 10-50 cm tall; Leaves simple, opposite, densely clustered near the base, linear lanceolate, base slightly connate, margin entire, thickened, apex acuminate, midrib thick, tri-or penta-nerved; Flowers bisexual, pedicellate, bracts 4, paired, ovate, spathulate-obovate, petals white, dark pink, purplish, base clawed, pubescent, staminodes absent, gynophore long, styles 2, filiform, stigmas 2; Fruit capsule, sub-sessile, ovoid-cylindrical, dehiscing by 4 teeth, carpophores present.

Flowering and Fruiting: June-September

136. Scientific Name: *Dichanthium annulatum* (Forssk.) Stapf

Synonym: *Andropogon annulatus* Forssk.

Family: Poaceae

Common Name: Kleberg's bluestem

Vernacular Name: Dilli Ghas

Habitat: Open tropical grasslands

Uses: Fodder, Erosion control

IUCN Threat listing: Not Assessed (NA)

Description: An erect or ascending, densely tufted perennial herb; Stem up to 1 m high, nodes bearded; Leaves linear-lanceolate, setaceous; Flowers in spiciform, usually greenish or purplish racemes; Spikelets sessile/pedicelled, bisexual, male or neuter.

Flowering and Fruiting: June-November

137. Scientific Name: *Dichrostachys cinerea* (L.) Wight & Arn.

Synonym: *Mimosa cinerea* L.

Family: Leguminosae

Common Name: Bell Mimosa

Vernacular Name: Kunali

Habitat: Open tropical areas

Uses: Avenue tree, Medicinal

IUCN Threat listing: Least Concerned (LC)

Description: Highly variable thorny shrub or small tree, up to 8 m tall, eaves bipinnate, 4-8 cm long, pinnae 5-15 pairs, each with 24-44 linear, obtuse to acute, straight to incurved leaflets; Inflorescence bi-coloured spike, upper pink flowers sterile with protruding staminodes, lower flowers yellow, bisexual, pistil 1, stamens 10, yellow; Pods dark-brown, twisted to form clusters, narrowly oblong, blackish, glabrous.

Flowering and Fruiting: October-January

138. Scientific Name: *Digitaria ciliaris* (Retz.) Koeler

Synonym: *Paspalum ciliare* (Retz.) DC.

Family: Poaceae

Common Name: Southern Crabgrass

Habitat: Open tropical grasslands

Uses: Fodder

IUCN Threat listing: Not Assessed (NA)

Description: An annual grass, typically decumbent, rooting at the nodes and spreading; Leaves 25 cm x 1 cm, sheaths and lower parts loosely hairy on both surfaces; The rachis of racemes up to 1 mm wide, spikelets in pairs, 1 sessile, other shortly pedicelled, lower glume very short, upper glume more than half the length of the spikelet, with three nerves.

Flowering and Fruiting: July-November

139. Scientific Name: *Digitaria eriantha* Steud.

Synonym: *Digitaria decumbens* Stent

Family: Poaceae

Common Name: Digit Grass

Habitat: Open tropical grasslands

Uses: Fodder

IUCN Threat listing: Not Assessed (NA)

Description: A perennial grass, culms simple or branched, 35 to 180 cm tall; Leaf blades 5-60 cm x 1 cm, glabrous or hairy, stolons spread over the surface of the ground and develop roots at the nodes; Inflorescence digitate or sub-digitate panicle, comprising 3-17 racemes, 5-20 cm long.

Flowering and Fruiting: June-October

140. Scientific Name: *Dinebra retroflexa* (Vahl) Panz.

Synonym: *Cynosurus retroflexus* Vahl

Family: Poaceae

Common Name: Viper Grass

Habitat: Open tropical grasslands

Uses: Fodder

IUCN Threat listing: Not Assessed (NA)

Description: A loosely tufted, annual, slender plant, usually straggling and ascending; Leaf-sheaths glandular, ligule membranous, leaf blade has an acuminate apex; Inflorescence linear with short oblong to cuneate, densely crowded, stiff racemes, more or less elliptic-oblong or pyramidal; spikelets 1-3 flowered, narrowly cuneate; Fruit an ellipsoid caryopsis with adherent pericarp.

Flowering and Fruiting: September-November

141. Scientific Name: *Dioscorea bulbifera* L.

Synonym: *Dioscorea heterophylla* Roxb.

Family: Dioscoreaceae

Common Name: Air Yam

Vernacular Name: Jimikand

Habitat: Open tropical areas

Uses: Edible, Medicinal

IUCN Threat listing: Not Assessed (NA)

Description: Herbaceous twiner, forming tubers; Stem glabrous, twining to the left, unarmed; Leaves opposite/alternate, ovate-cordate, membraneous, acuminate, bearing caducous bulbils in the axils; Male flowers white, in shorter spikes, female creamy, in longer spikes; Capsules quadrately-oblong.

Flowering and Fruiting: March-October

142. Scientific Name: *Dioscorea villosa* L.

Synonym: *Dioscorea sativa* L.

Family: Dioscoreaceae

Common Name: Wild Yam

Vernacular Name: Varah Kand

Habitat: Open tropical areas

Uses: Edible, Medicinal

IUCN Threat listing: Not Assessed (NA)

Description: A herbaceous vine, up to 10 m long, branching occasionally, stems slender, twining around vegetation, climbing upward and outward; Leaves whorled and/or opposite near the base, otherwise alternate; Plant dioecious, vines have either all male flowers or all female flowers, male vines have axillary panicles of numerous male flowers, female have axillary racemes of 5-15 flowers; Fruits flattened, broad membranous winged, distributed by wind.

Flowering and Fruiting: May-August

143. Scientific Name: *Diospyros melanoxylon* (Buch.-Ham.) V. Singh

Synonym: *Diospyros montana* B.Heyne *ex* A.DC.

Family: Ebenaceae

Common Name: Coromandel Ebony

Vernacular Name: Tendu

Habitat: Open tropical areas

Uses: Edible, Cigarette, Medicinal

IUCN Threat listing: Not Assessed (NA)

Description: A tall tree, up to length of 12-15 m; Leaves dark-green, simple, opposite, elliptic, entire and petiolate; Flowers solitary, campanulate and purple; Fruit berry, becoming pale yellow at maturity.

Flowering and Fruiting: April-July

144. Scientific Name: *Diplocyclos palmatus* (L.) C. Jeffrey

Synonym: *Bryonia palmata L.*

Family: Cucurbitaceae

Common Name: Lollipop climber

Vernacular Name: Shivlingi

Habitat: Open tropical areas

Uses: Edible, Medicinal

IUCN Threat listing: Not Assessed (NA)

Description: A perennial climbing herb, leafy stem glabrous, except few hairs in young parts; Inflorescences up to 10 male flowers of various ages, with up to 5 female flowers, or a few female flowers without male flowers; Fruits 1-5 in a cluster at the nodes, ripening bright red with (usually) pure white bands or rows of white blotches.

Flowering and Fruiting: September-January

145. Scientific Name: *Dombeya wallichii* (Lindl.) Benth. & Hook. *f.*

Synonym: *Assonia wallichii* (Lindl.) Kuntze

Family: Malvaceae

Common Name: Showy Dombeya

Vernacular Name: Domrupani

Habitat: Open tropical areas

Uses: Ornamental

IUCN Threat listing: Not Assessed (NA)

Description: A small tree, branching near base; Leaves heart-shaped, toothed, long-petiolate; Flowers in dense heads of little pink flowers, hanging from long stalks, showy, and fragrant.

Flowering and Fruiting: July-September

146. Scientific Name: *Duranta erecta* L.

Synonym: *Duranta repens* L.

Family: Verbenaceae

Common Name: Sky Flower

Vernacular Name: Nilkanta

Habitat: Open tropical areas

Uses: Ornamental, Hedge species, Medicinal

IUCN Threat listing: Not Assessed (NA)

Description: Large, evergreen shrub with multiple stems and hanging branches; Leaves ovate with serrate to entire leaf margin; Flower light purple, tubular, 5-lobed, in terminal or lateral clusters known as racemes; Fruits golden, round to slightly elongated berries.

Flowering and Fruiting: Throughout the Year

147. Scientific Name: *Echinochloa colona* (L.) Link

Synonym: *Panicum colonum* L.

Family: Poaceae

Common Name: Jungle Rice

Vernacular Name: Jharwa

Habitat: Open tropical grasslands

Uses: Fodder, Edible

IUCN Threat listing: Least Concerned (LC)

Description: Annual, branched/unbranched grass; Stem erect or ascending, up to 70 cm high; Leaves linear or lanceolate; Spikes 8-20, sessile, spikelets crowded, ovoid, purplish; Glumes ovate, membraneous, never awned, 3-4-seriate.

Flowering and Fruiting: Throughout the Year

148. Scientific Name: *Eclipta prostrata* (L.) L.

Synonym: *Eclipta alba* (L.) Hassk.

Family: Asteraceae

Common Name: False Daisy

Vernacular Name: Bhringraj

Habitat: Open tropical areas

Uses: Cosmetics, Medicinal

IUCN Threat listing: Data deficient

Description: Variable annual; Stems erect or prostrate, often rooting at the nodes; Leaves linear or oblong-lanceolate, narrowed at both ends, entire or serrate; Heads solitary or 2-3 togeher, on unequal peduncles, pale white; Achenes warty, pappus indistinct.

Flowering and Fruiting: Throughout the Year

149. Scientific Name: *Eichhornia crassipes* (Mart.) Solms

Synonym: *Pontederia crassipes* Mart.

Family: Pontederiaceae

Common Name: Water Hyacinth

Vernacular Name: Jal Kumbhi

Habitat: Humid tropical areas

Uses: Phytoremediation, Fibre, Manure

IUCN Threat listing: Not Assessed (NA)

Description: A perennial, free floating hydrophyte, stem floating by bouyant bulb-like nodules at its base, roots feathery, freely hanging; Leaves 10-20 cm across, above water; Flowers 8-15 in spike, on erect stalk, conspicuously attractive, lavender to pink, petals 6.

Flowering and Fruiting: June-November

150. Scientific Name: *Elaeocarpus ganitrus* Roxb.

Synonym: Elaeocarpus sphaericus (Gaertn.) K. Schum

Family: Elaeocarpaceae

Common Name: Blue Fig

Vernacular Name: Rudraksh

Habitat: Humid tropical areas

Uses: Sacred, Medicinal

IUCN Threat listing: Not Assessed (NA)

Description: An evergreen, erect tree, bark brownish, smooth, blaze orange red; Leaves simple, alternate, spiral, clustered at twig ends, stipules small, caduceus, petiole swollen at both ends; Flowers in racemes, petals white, laciniate, anthers ciliate; Drupe, oblong or ovoid, seeds 3-4.

Flowering and Fruiting: July-October

151. Scientific Name: *Elaeodendron glaucum* (Rottb.) Pers.

Synonym: *Cassine glauca* (Rottb.) Kuntze

Family: Celastraceae

Common Name: Ceylon Tea

Vernacular Name: Jamrasi

Habitat: Open tropical areas

Uses: Fuel wood, Timber, Medicinal

IUCN Threat listing: Data deficient

Description: A large tree, with spreading crown, bark red inside, exuding a watery sap when cut; Leaves opposite or sub-opposite, 3-16 x 1.5-7.5 cm, elliptic, ovate-oblong or obovate, glabrous, dark green, shining above, glaucous beneath; Flower whitish, calyx 4-5-lobed, lobes rounded, margin membranous, stamens included, filaments recurved; Fruit a dry, ovoid, drupe, 1-15 cm long, 1-celled, 1-seeded.

Flowering and Fruiting: February-November

152. Scientific Name: *Elephantopus scaber* L.

Synonym: *Elephantopus sordidus* Salisb.

Family: Asteraceae

Common Name: Elephant Foot

Vernacular Name: Samdudri

Habitat: Open tropical areas

Uses: Medicinal

IUCN Threat listing: Not Assessed (NA)

Description: An erect, annual herb height up to 40 cm; Leaves 12.5-20 cm long, mostly radical forming a spreading rosette on the ground, obovate-oblong, coarsely serrate-dentate, hairy; Heads numerous, sessile, closely packed, in large, flat-topped terminal inflorescence, flowers small, violet.

Flowering and Fruiting: October-January

153. Scientific Name: *Eleusine indica* (L.) Gaertn.

Synonym: *Cynodon indicus* (L.) Raspail

Family: Poaceae

Common Name: Indian Goosegrass

Vernacular Name: Balraj

Habitat: Open tropical areas

Uses: Edible, Medicinal

IUCN Threat listing: Least Concerned (LC)

Description: Erect annual, up to 60 cm high; Stem branched or unbranched, culms glabrous, somewhat compressed; Leaves distichous, sheath ciliate; Spikes digitate, 2-7, clustered at top, rachis flattened, spikelets biseriate, secund.

Flowering and Fruiting: July-January

154. Scientific Name: *Emilia sonchifolia* (L.) DC. *ex* Wight

Synonym: *Senecio sonchifolius* (L.) Moench

Family: Asteraceae

Common Name: Red Tasselflower

Vernacular Name: Hirankuri

Habitat: Open tropical grasslands

Uses: Edible, Medicinal

IUCN Threat listing: Not Assessed (NA)

Description: An erect, glabrous, milky annual; Stems erect or diffuse; Radical leaves spreading, usually stalked, entire or sinuate, upper ones small, few, alternate, stem clasping; Heads pink, solitary or laxly corymbose; Achenes 5-angled, angles bristly.

Flowering and Fruiting: November-March

155. Scientific Name: *Epiphyllum oxypetalum* (DC.) Haw.

Synonym: *Cactus oxypetalus Moc. & Sessé ex DC.*

Family: Cactaceae

Common Name: Dutchman's Pipe Cactus

Vernacular Name: Nishagandhi

Habitat: Humid tropical areas

Uses: Ornamental, Medicinal

IUCN Threat listing: Least Concerned (LC)

Description: An epiphytic herb, stems flattened, with a thick, round midrib surrounded by fleshy wings that have a scalloped edge; Flowers fragrant, star-shaped, tepals 25-30, white, linear, stamens greenish white, slender, styles greenish white, many lobed; Fruit oblong, purplish red, angled.

Flowering and Fruiting: June-October

157. Scientific Name: *Eragrostis amabilis* (L.) Wight & Arn

Synonym: *Eragrostis tenella* (L.) P.Beauv. ex Roem. & Schult.

Family: Poaceae

Common Name: Japanese Lovegrass

Habitat: Open tropical grasslands

Uses: Fodder

IUCN Threat listing: Not Assessed (NA)

Description: A delicate, erect, annual herb, forming loose tufts, stem glabrous, nodes dark; Leaf elongated, linear, with sharp pointed tip, glabrous, with scabrous margin; Inflorescence loose and branched, pyramidal to elliptical panicle, with very fine twigs provided with a tuft of hairs at the base, spikelets flattened, pedicellate, each with 4-10 florets, arranged sparsely.

Flowering and Fruiting: July-November

157. Scientific Name: *Eragrostis mino*r Host

Synonym: *Poa eragrostis* L.

Family: Poaceae

Common Name: Little Lovegrass

Habitat: Open tropical grasslands

Uses: Fodder

IUCN Threat listing: Not Assessed (NA)

Description: A loosely tufted grass; Leaf blades flat, mostly glabrous, usually with a row of warty glands along the margin; Spikelets 10-40-flowered, narrowly oblong or almost linear, yellowish green, breaking up from the base; Caryopsis broadly oblong, dark brown.

Flowering and Fruiting: May-September

158. Scientific Name: *Erigeron bonariensis* L.

Synonym: *Leptilon bonariense* (L.) Small

Family: Asteraceae

Common Name: Hairy Fleabane

Habitat: Open tropical areas

Uses: Antimicrobial, Medicinal

IUCN Threat listing: Not Assessed (NA)

Description: Annual, herbaceous weed, reaching up to 100 cm; Leaves linear to oblanceolate, entire, often wavy-edged, with very short or hooked hairs; Inflorescence long branched, thus corymbose, most flowering heads about the same level, Heads greyish-green, disc florets cream-coloured, ray florets absent; Pappus white or pinkish, seeds about 1 mm long.

Flowering and Fruiting: Throughout the Year

159. Scientific Name: *Eryngium foetidum* L.

Synonym: *Eryngium antihystericum* Rottb.

Family: Apiaceae

Common Name: Long Coriander

Vernacular Name: Ban Dhaniya

Habitat: Humid tropical areas

Uses: Edible, Medicinal

IUCN Threat listing: Not Assessed (NA)

Description: A perennial herb, 25-40 cm in height, stems dichotomously branched, deeply striate; Leaves simple, lamina oblanceolate-spathulate, cuneate at base, obtuse at apex; Flowers white, in oblong-cylindrical umbels, bracts spinulose, stellate-pubescent, calyx teeth rigid, acute, petals white; Fruit ellipsoid.

Flowering and Fruiting: Throughout the Year

160. Scientific Name: *Eucalyptus tereticornis* Sm.

Synonym: *Eucalyptus insignis* Naudin

Family: Myrtaceae

Common Name: Forest Red Gum

Vernacular Name: Nilgiri

Habitat: Open tropical areas

Uses: Fuel wood, Timber, Oil, Medicinal

IUCN Threat listing: Not Assessed (NA)

Description: A very large tree; Stem with grey bark, peeling in thin layers; Leaves alternate, linear-lanceolate; Flowers in umbels of 4-8, operculum conical, acuminate, longer than the calyx tube, long-pointed; Fruits globose, turbinate.

Flowering and Fruiting: June-November

161. Scientific Name: *Eugenia roxburghii* DC.

Synonym: *Syzygium bracteatum* (Willd.) Raizada

Family: Myrtaceae

Common Name: Roxburgh's Cherry

Vernacular Name: Chirai Jamun

Habitat: Humid tropical areas

Uses: Edible, Medicinal

IUCN Threat listing: Not Assessed (NA)

Description: A small tree with a spreading crown; Leaves ovate or elliptic-lanceolate, glabrous, caudate-acuminate, base acute; Flowers in axillary, 2 or 3-flowered cymes, bracteoles linear; Fruit a deep orange-colored berry.

Flowering and Fruiting: July-November

162. Scientific Name: *Eulaliopsis binata* (Retz.) C.E. Hubb.

Synonym: *Andropogon binatus* Retz.

Family: Poaceae

Common Name: Sabai Grass

Vernacular Name: Babui

Habitat: Open tropical grasslands

Uses: Paper making, Fiber, Fodder

IUCN Threat listing: Not Assessed (NA)

Description: Perennial grass; Stem woody, compressed, up to 50 cm tall; Leaves linear, ±15 cm long, basal crowded, white tomentose; Inflorescence of 2-4, digitate or fascicled racemes, golden or dark yellow pubescent; Spikelets sessile, lanceolate, glumes obtuse, ciliate in upper half.

Flowering and Fruiting: June-December

163. Scientific Name: *Euphorbia cotinifolia* L.

Synonym: *Alectoroctonum cotinifolium* (L.) Schltdl.

Family: Euphorbiaceae

Common Name: Red Spurge

Vernacular Name: Lal Dudhi Vriksh

Habitat: Humid tropical areas

Uses: Avenue tree, Fish poision, Medicinal

IUCN Threat listing: Not Assessed (NA)

Description: Shrubby to small tree, stems flexible, purplish, bark gray, smooth; Leaves coppery-purple, thin, fleshy, broad; Flowers inconspicuous, green or white, in cyathia, bracts creamy; Fruit brown capsule, very small.

Flowering and Fruiting: August-October

164. Scientific Name: *Euphorbia heterophylla* L.

Synonym: *Poinsettia heterophylla* (L.) Klotzsch & Garcke

Family: Euphorbiaceae

Common Name: Wild Poinsettia

Vernacular Name: Jangli Dudhi

Habitat: Open tropical areas

Uses: Medicine, Dye

IUCN Threat listing: Not Assessed (NA)

Description: Erect or ascending, stout, annual herb; Stems branched, fistular, ribbed; Leaves alternate at lower nodes, opposite below, variable in shape, hairy and whitish below; Floral leaves white towards the base, cyathia with 1, round nectary; Capsules glabrous.

Flowering and Fruiting: June-August

165. Scientific Name: *Euphorbia hirta* L.

Synonym: *Chamaesyce hirta* (L.) Millsp.

Family: Euphorbiaceae

Common Name: Garden Spurge

Vernacular Name: Dudhi

Habitat: Open tropical areas

Uses: Wound healing, Medicinal

IUCN Threat listing: Not Assessed (NA)

Description: Annual, erect or decumbent herb; Stem usually quadrangular, clothed with yellowish hairs; Leaves opposite, elliptic, serrate, hairy and dark green above, pubescent and pale green beneath; Cyathia in pedunculate, globose or sub-globose cymose heads; Fruit a capsule, depressed-globose, hairy, trigonous.

Flowering and Fruiting: Throughout the Year

166. Scientific Name: *Euphorbia milii* Des Moul.

Synonym: *Euphorbia breonii* Nois., *nom. illeg.*

Family: Euphorbiaceae

Common Name: Crown of Thorns

Habitat: Open tropical areas

Uses: Ornamental, Antimicrobial, Medicinal

IUCN Threat listing: Data deficient (DD)

Description: A succulent, diffusely branched shrub, up to 1.8 m high; Leaves obovate, broad, appear on new branches; Flowers small, subtended by petal-like bracts, variably red, pink or white.

Flowering and Fruiting: Throughout the Year

167. Scientific Name: *Euphorbia pulcherrima* Willd. *ex* Klotzsch

Synonym: *Poinsettia pulcherrima* (Willd. *ex* Klotzsch) Graham

Family: Euphorbiaceae

Common Name: Poinsettia

Habitat: Open tropical areas

Uses: Ornamental, Antimicrobial, Medicinal

IUCN Threat listing: Not Assessed (NA)

Description: A shrub or small tree, up to 4 m high; Leaves dark green, dentate, with red midrib and veins; Flowers greenish-yellow, inconspicuous, bracts attractive fire red, orange, pink, white or marbled, 10-20 surround few flower heads.

Flowering and Fruiting: December-March

168. Scientific Name: *Euphorbia thymifolia* L.

Synonym: *Euphorbia microphylla* L.

Family: Euphorbiaceae

Common Name: Little Spurge

Vernacular Name: Chhoti Dudhi

Habitat: Open tropical areas

Uses: Insect repellent, Medicinal

IUCN Threat listing: Not Assessed (NA)

Description: A small, slender, annual, prostrate herb, more or less hispidly pubescent; Leaves opposite, very small, obliquely oblong or elliptic-oblong, crenulate; Involucres of cyathia axillary, solitary, campanulate; Capsules minute, hairy.

Flowering and Fruiting: November-May

169. Scientific Name: *Evolvulus alsinoides* (L.) L.

Synonym: *Convolvulus alsinoides* L.

Family: Convolvulaceae

Common Name: Dwarf Morning Glory

Vernacular Name: Vishnukranta

Habitat: Open tropical areas

Uses: Aromatic, Medicinal

IUCN Threat listing: Not Assessed (NA)

Description: Diffuse, perennial, prostrate herb; Stems appressed hairy, trailing, much branched; Leaves 1-1.5 x 0.3-0.5 cm, lanceolate to ovate, pilose, apex mucronate; Flowers blue or white, axillary, 1-3-together, peduncled; Capsules globose, 4-valved, usually 4-seeded.

Flowering and Fruiting: July-November

170. Scientific Name: *Evolvulus nummularius* (L.) L.

Synonym: *Convolvulus nummularius* L.

Family: Convolvulaceae

Common Name: Roundleaf Bindweed

Vernacular Name: Musakarni

Habitat: Open tropical areas

Uses: Medicinal

IUCN Threat listing: Not Assessed (NA)

Description: Prostrate, creeping, annual herbs; Stem slender; Leaves glabrous, shining, apex obtuse or retuse; Flowers white, solitary, axillary or in cymes; Capsules globose.

Flowering and Fruiting: Throughout the Year

171. Scientific Name: *Ficus benghalensis* L.

Synonym: *Ficus banyana* Oken

Family: Moraceae

Common Name: Banyan

Vernacular Name: Barr

Habitat: Humid tropical areas

Uses: Edible, Medicinal, Latex, Wound healing

IUCN Threat listing: Not Assessed (NA)

Description: Large, evergreen tree; Stem with greyish-white, smooth bark, crown dense, branches horizontally spreading, giving down aerial roots; Leaves coriaceous, alternate, ovate/elliptic, obtuse, entire, base sub-cordate or rounded, petiole stout; Receptacles in axillary pairs, globose, sessile, red when ripe; Syconia nearly 1 cm in diameter, supported by 3 broad, basal bracts.

Flowering and Fruiting: May-August

172. Scientific Name: *Ficus elastica* Roxb.

Synonym: *Ficus clusiifolia* Summerhayes

Family: Moraceae

Common Name: Rubber Fig

Vernacular Name: Shimal

Habitat: Humid tropical areas

Uses: Edible, Ornamental, Medicinal, Latex

IUCN Threat listing: Not Assessed (NA)

Description: A large, evergreen, glabrous tree, up to 30 m high, crown dense, branches spreading; Leaves long petiolate, lamina thick, glossy above, oblong or elliptic; Hypanthodia sessile, in extra axillary pairs, pale-green, subtended by 3 caducous bracts, male flowers dispersed in receptacle, female flowers sessile, ovary smooth; Figs ovoid-oblong, 10-12 mm long, 6-8 mm across, pale to yellowish brown.

Flowering and Fruiting: November-January

173. Scientific Name: *Ficus hispida* L. *f.*

Synonym: *Ficus scabra* Jacq.

Family: Moraceae

Common Name: Hairy Fig

Vernacular Name: Kat Gular

Habitat: Humid tropical areas

Uses: Edible, Medicinal

IUCN Threat listing: Not Assessed (NA)

Description: A small tree of tropical forests; Leaves opposite, stalks 1-4 cm, lamina ovate, oblong, thickly papery, covered with coarse hairs; Figs solitary or paired, in leaf axils on leafy shoots or sometimes leafless branchlets, yellow or red when mature.

Flowering and Fruiting: September-May

174. Scientific Name: *Ficus mollis* Vahl

Synonym: *Ficus pubescens* Heyne *ex* Roth

Family: Moraceae

Common Name: Soft Fig

Habitat: Humid tropical areas

Uses: Medicinal

IUCN Threat listing: Not Assessed (NA)

Description: Deciduous tree, epiphytic or independent, up to 15 m high; Leaves simple, alternate, spiral, stipule lateral, ovate-lanceolate, caducous, leaving an annular scar; Syconia sessile, in axillary pairs, sub-globose, tomentose, orifice circular, slightly raised, closed by 3 glabrous apical bracts, flowers unisexual; Figs 5-8 mm across, fleshy, grey, tomentose, achenes smooth.

Flowering and Fruiting: March-April

175. Scientific Name: *Ficus palmata* Forssk.

Synonym: *Ficus gossypina* Wall.

Family: Moraceae

Common Name: Punjab Fig

Vernacular Name: Anjiri

Habitat: Humid tropical areas

Uses: Edible, Medicinal

IUCN Threat listing: Not Assessed (NA)

Description: Bushy to small tree, up to 10 m, bark smooth, brownish-grey, young twigs hairy; Leaves broadly suborbicular, orbicular to ovate, cordate at the base, petiole tomentose, mostly undivided or sometimes palmate, partite, dentate to serrate, acute to obtuse; Male flowers numerous in upper half, pedicellate, sepals 5, free, stamens 3-6, female flowers basal, numerous, sepals 5, basally united; Figs gradually narrowed at the base, 2 cm long, yellow or purple in color.

Flowering and Fruiting: March-July

176. Scientific Name: *Ficus racemosa* L.

Synonym: *Ficus glomerata* Roxb.

Family: Moraceae

Common Name: Cluster Fig

Vernacular Name: Gular

Habitat: Humid tropical areas

Uses: Edible, Medicinal, Erosion control

IUCN Threat listing: Not Assessed (NA)

Description: Large tree; Stem with smooth, reddish-brown bark; Leaves alternate, membraneous, elliptic-lanceolate, entire, glabrous on upper surface, sub-acute; Receptacles on short warted, leafless branchlets; Syconia peduncled, sub-globose, 2-4 cm in diameter, downy, reddish-orange on ripening.

Flowering and Fruiting: February-May

177. Scientific Name: *Ficus religiosa* L.

Synonym: *Ficus peepul* Griff.

Family: Moraceae

Common Name: Holy Fig

Vernacular Name: Peepal

Habitat: Open tropical areas

Uses: Edible, Gum, Medicinal, Dye

IUCN Threat listing: Not Assessed (NA)

Description: Large, deciduous tree, epiphytic in early life; Trunk irregularly-shaped, bark whitish-grey bearing brown specks; Leaves entire, glabrous, shining, reddish when young, cordate, with 5-7 basal nerves, sharply acuminate; Receptacles in axillary pairs, sessile, smooth; Syconia depressed-globose, 1.5 cm in diameter, dark purple on ripening.

Flowering and Fruiting: November-February

178. Scientific Name: *Ficus semicordata* Buch. *ex* J.E. Smith

Synonym: *Ficus cunia* Buch.-Ham. *ex* Roxb.

Family: Moraceae

Common Name: Drooping Fig

Vernacular Name: Ghui

Habitat: Humid tropical areas

Uses: Edible, Medicinal, Fibre, Slope stabilization

IUCN Threat listing: Not Assessed (NA)

Description: Moderate-sized tree; Stem with dark grey bark and scabrous branches; Leaves alternate, elliptic to oblong-lanceolate, distinctly unequal-sided, acuminate, crenate-serrate; Receptacles globose or pyriform, in pairs or clusters on leafless shoots; Syconia ribbed and brown on ripening.

Flowering and Fruiting: July-September

179. Scientific Name: *Fimbristylis littoralis* Gaudich.

Synonym: *Fimbristylis flaccida* Steud.

Family: Cyperaceae

Common Name: Grass like Fimbristylis

Habitat: Humid tropical areas

Uses: Fodder

IUCN Threat listing: Least Concerned (LC)

Description: A tufted, leafy, annual or short-lived, aquatic herb; Leaves up to 40 cm long, basal leaves half as long as culm, linear, threadlike and stiff, 2-ranked, with sheaths; Inflorescence a lax and diffuse compound umbel, 6-10 cm long, spikelets globose or ovate; Achenes obovoid, trigonous, biconvex, apiculate at apex, pale ivory to brown.

Flowering and Fruiting: April-July

180. Scientific Name: *Fimbristylis quinquangularis* Tang & F.T.Wang

Synonym: *Fimbristylis miliacea* (L.) Vahl

Family: Cyperaceae

Common Name: Five-angle Fimbri

Vernacular Name: Dilli

Habitat: Humid tropical areas

Uses: Fodder

IUCN Threat listing: Least Concerned

Description: Erect, tufted, non-rhizomatous annuals, 10-40 cm tall; Leaves basal, equitant, 10-35 cm long, linear, laterally flattened, scabrid on margins; sheaths 2-8 cm long, laterally compressed, ligule absent; Spikelets solitary, globose, obtuse, rusty-brown, rachilla wingless, glumes spiral.

Flowering and Fruiting: Throughout the Year

181. Scientific Name: *Flemingia macrophylla* (Willd.) Merr.

Synonym: *Flemingia semialata* W.T. Aiton

Family: Leguminosae

Common Name: Large Leaf Flemingia

Habitat: Humid tropical areas

Uses: Hedge plant, Mulching, Fodder, Medicinal

IUCN Threat listing: Not Assessed (NA)

Description: An erect undershrub; Stem 1.2-1.8 m high, with slightly ribbed branches; Leaves 3-foliolate, petiole flattened, leaflets ovate/oblong-lanceolate, glabrous above, hairy below; Flowers in dense, sessile, 2-5 cm long racemes; Pods 1-2 cm long, downy, turgid, hairy.

Flowering and Fruiting: November-January

182. Scientific Name: *Gaillardia aristata* Pursh

Synonym: *Gaillardia bicolor* Pursh

Family: Asteraceae

Common Name: Blanket flower

Habitat: Open tropical areas

Uses: Ornamental

IUCN Threat listing: Not Assessed (NA)

Description: A perennial herb, 20-70 cm in height; Leaves lance-shaped, mainly basal, upper stems naked; Capitula have brown to reddish purple disc, ray florets yellow to reddish with dark bases; Fruit a stout, hairy achene, may be more than 1 cm long.

Flowering and Fruiting: Throughout the Year

183. Scientific Name: *Galphimia speciosa* C.E. Anderson

Synonym: *Galphimia glauca* Hort. *ex* Bartl.

Family: Malpighiaceae

Common Name: Gold Shower

Habitat: Open tropical areas

Uses: Ornamental

IUCN Threat listing: Not Assessed (NA)

Description: A robust, evergreen, ornamental shrub, up to 2 m high, trunk covered with reddish fibres; Leaves alternate, thin, leathery, greyish green in colour; Flowers yellow, pentamerous, stamens 10, styles 3.

Flowering and Fruiting: October-December

184. Scientific Name: *Gardenia jasminoides* J. Ellis

Synonym: *Gardenia angustifolia* G.Lodd.

Family: Rubiaceae

Common Name: Cape Jasmine

Habitat: Open tropical areas

Uses: Ornamental, Dye, Aromatic, Medicinal

IUCN Threat listing: Not Assessed (NA)

Description: Small, ornamental shrub, up to 3 m in height, twigs often short-haired, cylindrical, gray; Leaves opposite, or 3 whorled, leathery, thin, lanceolate, oblong, obovate or elliptic; Flowers fragrant, usually 1-flowered on branchlets, pedicels obconical or ovate; Fruits ovate, sub-globose, elliptic or oblong, yellow or orange-red.

Flowering and Fruiting: June-October

185. Scientific Name: *Gardenia resinifera* Roth

Synonym: *Gardenia lucida* Roxb.

Family: Rubiaceae

Common Name: Brilliant Gardenia

Habitat: Open tropical areas

Uses: Insect repellent, Medicinal, Gum

IUCN Threat listing: Not Assessed (NA)

Description: A woody shrub, with attractive foliage and fragrance; Leaves in whorls of 3, green, glossy, margin entire, about 10 cm long; Flowers solitary, fragrant and white; Fruits small, oval, green when young, orange when ripe.

Flowering and Fruiting: September-January

186. Scientific Name: *Garuga pinnata* Roxb.

Synonym: *Kunthia pinnata* (Roxb.) Kuntze

Family: Burseraceae

Common Name: Grey Downy Balsam

Vernacular Name: Kekad

Habitat: Open tropical areas

Uses: Edible

IUCN Threat listing: Not Assessed (NA)

Description: Large, deciduous tree; Stem with greyish-brown bark; Leaves 20-35 cm long, exstipulate, leaflets 5-9 pairs, opposite, ovate-lanceolate, crenate, base rounded; Flowers polygamous, in large terminal panicles, yellow; Drupes globose, 1.5-2.0 cm in diameter.

Flowering and Fruiting: January-July

187. Scientific Name: *Glinus oppositifolia* (L.) A. DC.

Synonym: *Mollugo spergula* L.

Family: Molluginaceae

Common Name: Indian Chickweed

Habitat: Open tropical areas

Uses: Edible, Medicinal

IUCN Threat listing: Not Assessed (NA)

Description: A diffuse or prostrate, glabrous herb; Leaves in whorls of 3-5, sub-succulent, sub-sessile, unequal, obovate-spathulate, base attenuate, apex obtuse or acute; Flowers in axillary, 5-8-flowered fascicles, tepals 5, oblong, with green central stripe, spreading at anthesis, closing again around fruit.

Flowering and Fruiting: February-April

188. Scientific Name: *Gloriosa superba* L.

Synonym: *Eugone superba* (L.) Salisb.

Family: Colchicaceae

Common Name: Flame Lily

Habitat: Open tropical areas

Uses: Ornamental, Medicinal

IUCN Threat listing: Least Concerned (LC)

Description: A herbaceous climber, with tuberous roots; Leaves sub-sessile, alternate, opposite or whorled, ovate-lanceolate, base cordate or amplexicaul, apex acuminate, ending in a tendril; Flowers bisexual, showy, axillary, solitary or in few-flowered, terminal racemes; Capsule 3-5 x 1-2 cm, ellipsoid-oblong, seeds many, globose, warty.

Flowering and Fruiting: July-December

189. Scientific Name: *Gmelina arborea* Roxb. *ex* Sm.

Synonym: *Premna arborea* Roth

Family: Lamiaceae

Common Name: Beechwood

Habitat: Open tropical areas

Uses: Fuel wood, Timber, Furniture

IUCN Threat listing: Not Assessed (NA)

Description: A deciduous tree, up to 15 m in length, bark white or whitish-grey, smooth, lenticellate, scurfy; Leaves simple, opposite, estipulate, petioles slender, tomentose, lamina broadly ovate or ovate; Flowers in terminal panicles, bisexual, yellow with orange-pink shade, campanulate, sepals tomentose, corolla bilabiate, densely hairy, stamens 4, didynamous; Fruit an ovoid, pulpy drupe with a deep depression, pyrene bonny.

Flowering and Fruiting: February-June

190. Scientific Name: *Gomphrena celosioides* Mart.

Synonym: *Gomphrena alba* Peter

Family: Amaranthaceae

Common Name: Bachelor's Button

Habitat: Open tropical areas

Uses: Antimicrobial, Medicinal

IUCN Threat listing: Not Assessed (NA)

Description: Perennial, decumbent herb; Stem hairy, prostrate, purplish green; Leaves opposite, sub-sessile or shortly-petioled, oblanceolate/spatulate, apex acute; Flowers in sessile heads, at first globose, gradually elongating into a wide spike; Fruit single-seeded urticles.

Flowering and Fruiting: September-December

191. Scientific Name: *Gossypium arboreum* L.

Synonym: *Hibiscus purpurascens* Kuntze

Family: Malvaceae

Common Name: Tree Cotton

Habitat: Open tropical areas

Uses: Edible, Medicinal, Fibre

IUCN Threat listing: Not Assessed (NA)

Description: A short shrub, up to 3 m, young branches purple pubescent with stellate and few simple hairs; Leaves simple, alternate, spiral, stipules linear-lanceolate, lamina palmately lobed to palmatifid; Flowers solitary, axillary or in terminal racemes, epicalyx segments large ovate, calyx persistent, corolla pink or white; Capsules depressed globose, pubescent, enclosed by calyx.

Flowering and Fruiting: August-December

192. Scientific Name: *Grangea maderaspatana* (L.) Poir.

Synonym: *Artemisia maderaspatana* L.

Family: Asteraceae

Common Name: Madras Carpet

Habitat: Open tropical areas

Uses: Edible, Medicinal

IUCN Threat listing: Least Concerned (LC)

Description: Annual, glandular-pubescent herb; Leaves radical, tapering at base, lyrate-pinnatifid, pubescent, lateral segments oblong, pair opposite or sub-opposite, obtuse, terminal segments oblong, trilobed; Capitula solitary or 2-4, terminal or axillary, involucral bracts oblong, spathulate, margins scarious, florets glandular.

Flowering and Fruiting: March-July

193. Scientific Name: *Grevillea robusta* A. Cunn. *ex* R. Br.

Synonym: *Stylurus robusta* (A. Cunn.) Deg.

Family: Proteaceae

Common Name: Silver Oak

Habitat: Open tropical areas

Uses: Dye, Furniture

IUCN Threat listing: Not Assessed (NA)

Description: Large, robust, oak-like tree; Main trunk huge with hoary or rusty tomentose young branches; Leaves alternate, pinnate, pinnae pinnatifid or entire, segments lanceolate, margins recurved, silvery beneath; Flowers orange-coloured, in secund, unilateral, showy racemes wih slender panicles; Fruit an oblique, coriaceous follicle, brown in colour.

Flowering and Fruiting: May-July

194. Scientific Name: *Grewia hirsuta* Vahl

Synonym: *Grewia longifolia* Buch.-Ham.

Family: Malvaceae

Common Name: Hairy Grewia

Habitat: Humid tropical areas

Uses: Medicinal, Fibre

IUCN Threat listing: Not Assessed (NA)

Description: A shrub or small tree, can grow up to 5 m tall; Leaves 6 x 2.5 cm, elliptic-lanceolate, finely serrate, obtuse apex, oblique, puberulus above, densely tomentose below; Cymes axillary, peduncles 1-3, tomentose, buds globose, tomentose, bracts and bracteoles lanceolate, sepals oblong, tomentose, petals ovate, pubescent below.

Flowering and Fruiting: August-December

195. Scientific Name: *Grewia oppositifolia* Buch.-Ham. *ex* D. Don

Synonym: *Grewia emarginata* Wight & Arn.

Family: Malvaceae

Common Name: Opposite Leaved Grewia

Habitat: Humid tropical areas

Uses: Edible, Fibre

IUCN Threat listing: Not Assessed (NA)

Description: Moderate-sized tree; Stem with whitish bark; Leaves broadly ovate-lanceolate, acuminate, serrate, pubescent below, unequally round at the base; Flowers in leaf-opposed umbellate cymes; Fruit a drupe, 1-4-lobed, fleshy, green, black when ripe.

Flowering and Fruiting: February-June

196. Scientific Name: *Guizotia abyssinica* (L. *f.*) Cass.

Synonym: *Bidens ramtilla* Wall. *ex* DC.

Family: Asteraceae

Common Name: Niger Seed

Habitat: Open tropical areas

Uses: Edible, Medicinal, Oil

IUCN Threat listing: Not Assessed (NA)

Description: Herbaceous annual, 0.5–1.5 m tall; Leaves opposite, sessile, sub-cordate to ovate-lanceolate, serrate, up to 22 cm long; Flowers in solitary or clustered heads, arranged in corymbs, to 2 cm across, yellow, involucre with ovate, biseriate scales, heads with 40–60 tubular hermaphroditic florets.

Flowering and Fruiting: August-October

197. Scientific Name: *Habenaria marginata* Colebr.

Synonym: *Orchis clavata* B.Heyne ex Hook.f.

Family: Orchidaceae

Common Name: Golden Yellow Habenaria

Habitat: Humid tropical areas

Uses: Ornamental

IUCN Threat listing: Not Assessed (NA)

Description: Annual herb, with tuberoids, 10-25 cm tall, stem reduced, with 3-5 closely crowded sub-basal leaves; Leaves sessile, spreading, and ovate to oblong, acute, with pale yellow margins; Flowers green and yellow, sepals green, dorsal broadly ovate, forming a hood with the petals, lateral sepals narrower, oblong-lanceolate.

Flowering and Fruiting: July-November

198. Scientific Name: *Habenaria roxburghii* Nicolson

Synonym: *Orchis plantaginea* Roxb.

Family: Orchidaceae

Common Name: Roxburgh's Habenaria

Habitat: Humid tropical areas

Uses: Ornamental

IUCN Threat listing: Not Assessed (NA)

Description: A tuberous, terrestrial herb, 25–35 cm tall; Leaves 2 or 3, at the base, orbicular, acute at apex, base sheathing; Flowers white, bracts 2 cm, lanceolate, sepals broadly ovate, lip 3-lobed, side lobes smaller, oblong, spur 4 cm, oblong, clavate at apex.

Flowering and Fruiting: August-October

199. Scientific Name: *Hallenia speciosa* (J. Koenig) S.R. Dutta

Synonym: *Costus speciosus* (J. Koenig) Sm.

Family: Costaceae

Common Name: Crepe Ginger

Habitat: Humid tropical areas

Uses: Ornamental, Medicinal

IUCN Threat listing: Not Assessed (NA)

Description: Annual (underground parts perennial), erect herb; Stem base slightly woody, apex twisted, up to 1 m high; Leaves spirally arranged, sub-sessile, oblanceolate/elliptic, glabrous, apex acute, base cuneate; Flowers in terminal globose spikes, white, with a yellow centre and pink bracts; Capsule globose-trigonous, bright red, many seeded.

Flowering and Fruiting: October-December

200. Scientific Name: *Hedychium coronarium* J.Koenig

Synonym: *Amomum filiforme* W.Hunter

Family: Zingiberaceae

Common Name: Ginger Lily

Habitat: Humid tropical areas

Uses: Edible, Medicinal

IUCN Threat listing: Not Assessed (NA)

Description: Herbaceous, upright plants, growing 1-2.5 m high, culms cylindrical, vigorous, unbranched, reddish at base, mostly covered by leaf sheets, one for each internode; Leaves 2-ranked, alternate, ligules membraneous, lamina oblong to lanceolate, narrowed to a slender point; Inflorescence spike, 2-3 flowers emerge per bract, calyx glabrous, corolla tube longer hidden between the bracts, opening obliquely; Fruit an oblong capsule, 2.5-3.5 cm long, dehiscing loculicidally by 3 valves.

Flowering and Fruiting: July-November

201. Scientific Name: *Helianthus strumosus* L.

Synonym: *Helianthus montanus* E.Watson

Family: Asteraceae

Common Name: Pale leaf Woodland Sunflower

Habitat: Open tropical areas

Uses: Ornamental

IUCN Threat listing: Least Concerned (LC)

Description: An exotic, erect, perennial, grows from 3 to 8 feet in height; Leaves opposite on lower part, thick, rough, main veins 3, white tomentose beneath; Heads heterogamous, disc florets many, bisexual, corolla tube yellow, stamens 5, tightly surrounding a single style which has a bifurcated tip; Fruit dry cypsela, 4-5.5 mm long, with two small awns.

Flowering and Fruiting: August-October

202. Scientific Name: *Helicteres isora* L.

Synonym: *Isora corylifolia* Schott & Endl.

Family: Malvaceae

Common Name: Indian Screw Tree

Vernacular Name: Marod Phali

Habitat: Open tropical areas

Uses: Antimicrobial, Medicinal

IUCN Threat listing: Not Assessed (NA)

Description: Shrub; Stem with grey bark, thin spreading branches, covered with hairs; Leaves bifarious, alternate, shortly-petioled, broad-obovate, obliquely cordate, serrate, tomentose; Flowers 2-5 cm long, orange-red in colour, in axillary clusters of 2-4; Fruit a capsule, 4-5 cm long, pubescent, consisting of five cylindrical, spirally twisted carpels on a long gynophore.

Flowering and Fruiting: April-December

203. Scientific Name: *Heliotropium supinum* L.

Synonym: *Piptoclaina malabarica* (Retz.) G. Don

Family: Boraginaceae

Common Name: Dwarf Heliotrope

Habitat: Open tropical areas

Uses: Medicinal

IUCN Threat listing: Not Assessed (NA)

Description: Decumbent or procumbent annual, mat forming herb, 4–50 cm long, stems densely silvery pubescent, spreading pilose; Leaf-blades grey-green, elliptic, obovate or ovate- lanceolate, rounded-acuminate at apex, cuneate at base; Flowers sessile, in bilateral scorpioid cymes, white; Fruits of 1-4 nutlets, usually only 1 by abortion, calyx accrescent, corolla quite durable.

Flowering and Fruiting: March-June

204. Scientific Name: *Hemerocallis fulva* (L.) L.

Synonym: *Hemerocallis crocea* Lam., *nom. superfl.*

Family: Xanthorrhoeaceae

Common Name: Orange Day lily

Habitat: Open tropical areas

Uses: Ornamental

IUCN Threat listing: Not Assessed (NA)

Description: Herbaceous perennial, growing from tuberous roots, stems 40-150 cm; Leaves linear, 50-90 cm x 1–3 cm; Flowers 5-12 cm across, orange-red, with a pale central line on each tepal; Fruit 3-valved capsule.

Flowering and Fruiting: June-August

205. Scientific Name: *Hibiscus mutabilis* L.

Synonym: *Abelmoschus mutabilis* (L.) Wall. *ex* Hassk.

Family: Malvaceae

Common Name: Cotton Rose mallow

Habitat: Open tropical areas

Uses: Edible, Medicinal, Fiber

IUCN Threat listing: Not Assessed (NA)

Description: Bushy shrubs or small trees, about 2-6 m tall, stems woody, branchlets densely pubescent with stellate and simple hairs; Leaves simple, alternate, broadly ovate to suborbicular, 7-11 veined, 3-7 palmate, margins coarsely dentate; Flowers bisexual, 5-10 cm across, pedicels slender, epicalyx 8-12, linear lanceolate, corolla showy and large, yellow, white, pinkish, campanulate.

Flowering and Fruiting: All year round

206. Scientific Name: *Hibiscus rosa-sinensis* L.

Synonym: *Abelmoschus genevii* Walp.

Family: Malvaceae

Common Name: China Rose

Habitat: Open tropical areas

Uses: Ornamental, Medicinal

IUCN Threat listing: Not Assessed (NA)

Description: A bushy, evergreen, cultivated shrub or small tree, 2.5–5 m high; Leaves ovate-lanceolate, more or less acuminate, irregularly and coarsely serrate towards the top; Flowers large, axillary solitary, corolla tubular below, red, stamens numerous; Fruit a capsule, not usually seen in cultivation.

Flowering and Fruiting: May-October

207. Scientific Name: *Hibiscus sabdariffa* L.

Synonym: *Abelmoschus cruentus* Bertol.

Family: Malvaceae

Common Name: Roselle

Habitat: Open tropical areas

Uses: Edible, Medicinal

IUCN Threat listing: Not Assessed (NA)

Description: Unarmed, erect shrubs, up to 3 m high, stem glabrous or sparsely pubescent; Leaves polymorphic, entire or palmately 3-5 fid, partite or lobed, midlobe longest, cuneate at base, serrate at margin, acute at apex, glabrous Flowers axillary, solitary or in racemose panicles, pedicels glabrous, jointed near base, epicalyx lobes 8-12, adnate to sepals, calyx lobes lanceolate to oblong-elliptic, green or purple, persistent, corolla orange, streaked with deep purple; Capsules ovoid, ca 1.5 cm across, densely strigose, seeds reniform.

Flowering and Fruiting: August-January

208. Scientific Name: *Hibiscus syriacus* L.

Synonym: *Hibiscus chinensis* DC.

Family: Malvaceae

Common Name: Rose of Sharon

Habitat: Open tropical areas

Uses: Edible, Ornamental, Medicinal

IUCN Threat listing: Not Assessed (NA)

Description: Bushy, deciduous shrubs, 2-6 m tall, stems woody, erect, branchlets densely pubescent with stellate hairs, becoming glabrous ultimately; Leaves simple, alternate, palmilobed, broadly ovate-deltoid to rhomboid-ovate, margins coarsely dentate; Flowers bisexual, pedicels slender, stellate pubescent, epicalyx 6-8, base connate, sepals 5, distinctly nerved, petals purple; Fruit a capsule, globose-cylindrical, apex obtuse, abruptly beaked.

Flowering and Fruiting: August-September

209. Scientific Name: *Hydrangea macrophylla* f. *normalis* (E.H. Wilson) H. Hara

Synonym: *Hortensia japonica* J.F. Gmel.

Family: Hydrangeaceae

Common Name: Big leaf Hydrangea

Habitat: Humid tropical areas

Uses: Ornamental, Phytoremediation, Medicinal

IUCN Threat listing: Not Assessed (NA)

Description: A deciduous shrub, growing to 2 m; Leaves opposite, simple, stalked, toothed and sometimes lobed; flowers in terminal, round or umbrella-shaped clusters, white, pink, or blue, 4- 5-parted, the sterile flowers are around the margins and are much enlarged.

Flowering and Fruiting: July-December

210. Scientific Name: *Hygrophila schulli* (Buch.-Ham.) M.R. Almeida & S.M. Almeida

Synonym: *Hygrophila auriculata* (Schumach.) Heine

Family: Acanthaceae

Common Name: Marsh Barbel

Habitat: Humid tropical areas

Uses: Medicinal

IUCN Threat listing: Least Concerned (LC)

Description: Stout, erect, roughly hairy herb; Stem up to 1 m high; Leaves sub-sessile, whorled, oblong-lanceolate, hairy beneath, bear a sharp, straight, yellow spine in the axil; Flowers whorled at each node, bluish or purplish; Capsules oblong, 1.5 x 0.4 cm, pointed.

Flowering and Fruiting: June-January

211. Scientific Name: *Impatiens balsamina* L.

Synonym: *Balsamina balsamina* Huth

Family: Balsaminaceae

Common Name: Garden Balsam

Habitat: Humid tropical areas

Uses: Ornmanental, Edible, Medicinal, Dye

IUCN Threat listing: Not Assessed (NA)

Description: An ornamental, annual herb, 30-90 cm high; Leaves up to 15 cm long, lanceolate, acuminate, deeply serrate; Flowers large, axillary, showy, of various colours Capsules spindle-shaped, tomentose.

Flowering and Fruiting: March-October

212. Scientific Name: *Indigofera astragalina* DC.

Synonym: *Indigofera karaiyarensis* Rajakumar

Family: Leguminosae

Common Name: Silky Indigo

Habitat: Open tropical grasslands

Uses: Medicinal

IUCN Threat listing: Not Assessed (NA)

Description: Short, sub-erect, undershrub or a bushy herb; Stem 4-angled, densely brown pubescent; Leaves imparipinnte, leaflets 5-11, opposite, oblong-obovate, hairy beneath, terminal leaflet longer; Flowers red, arranged in 5-10 cm long, very dense, erect racemes; Pods 1.5-2.5 x 0.2-0.4 cm, densely hairy.

Flowering and Fruiting: August-November

213. Scientific Name: *Indigofera hirsuta* L.

Synonym: *Anila hirsuta* (L.)Kuntze

Family: Leguminosae

Common Name: Hairy Indigo

Habitat: Open tropical grasslands

Uses: Dye, Medicinal, Erosion control

IUCN Threat listing: Not Assessed (NA)

Description: An erect or spreading annual, up to 1 .5 m tall, stems cylindrical or slightly ridged, densely clothed with grey or reddish-brown pubescence; Leaves pinnately compound, leaflets hirsute; Inflorescence a dense, many-flowered hirsute raceme, calyx stiff, brown and hirsute, corolla white, pubescent outside, brick-red or rose inside; Pods straight, rather tetragonal, with well-developed sutures.

Flowering and Fruiting: September-February

214. Scientific Name: *Indigofera linifolia* (L.*f.*) Retz.

Synonym: *Hedysarum linifolium* L.*f.*

Family: Leguminosae

Common Name: Narrow Leaf Indigo

Habitat: Open tropical grasslands

Uses: Edible

IUCN Threat listing: Least Concerned (LC)

Description: A trailing or procumbent herb; Stem much branched, silvery-white; Leaves linear, sessile or nearly so, acute at both ends, mucronate; Flowers bright red, in 6-12-flowered, very short, axillary racemes; Pods small, silvery-white, globose or round, 1-seeded.

Flowering and Fruiting: June-September

215. Scientific Name: *Ipomoea aquatica* Forssk.

Synonym: *Convolvulus repens* Vahl

Family: Convolvulaceae

Common Name: Swamp Morning Glory

Habitat: Humid tropical areas

Uses: Edible, Medicinal

IUCN Threat listing: Least Concerned (LC)

Description: A sprawling, annual or perennial, creeping vine, stems terete, branched, hollow and succulent when floating, otherwise solid and firm, up to 3 m long; Leaves alternate, glabrous, petioles succulent, blades greenish-brown, triangular, ovate, lanceolate, entire to dentate; Flowers large and showy, corolla funnel shaped, glabrous, pink, often with darker eye; Fruit a capsule, glabrous, globose to ovoid, seeds up to 4.

Flowering and Fruiting: All year round

216. Scientific Name: *Ipomoea carnea* Jace.

Synonym: *Convolvulus batatilla* Kunth

Family: Convolvulaceae

Common Name: Bush Morning Glory

Habitat: Humid tropical areas

Uses: Edible, Medicinal, Erosion control

IUCN Threat listing: Not Assessed (NA)

Description: Water-loving shrub; Stem erect or diffuse, branches fistulose with milky latex; Leaves ovate/obovate, glabrous, apex acute-acuminate, base hastate or cordate; Flowers in lax cymes, pale pink Capsules ovoid.

Flowering and Fruiting: March-June

218. Scientific Name: *Ipomoea eriocarpa* R. Br.

Synonym: *Convolvulus angulatus* Spreng.

Family: Convolvulaceae

Common Name: Tiny Morning Glory

Habitat: Open tropical areas

Uses: Edible, Medicinal

IUCN Threat listing: Not Assessed (NA)

Description: Twining, annual, hairy herb; Stem slender; Leaves petioled, ovate-oblong, apex acute-acuminate, base hastate cordate; Flowers in sub-sessile heads, pink, ovary very hairy; Capsule globose, 0.6 cm in diameter.

Flowering and Fruiting: October-January

218. Scientific Name: *Ipomoea hederifolia* L.

Synonym: *Convolvulus sanguineus* (Vahl) Spreng.

Family: Convolvulaceae

Common Name: Scarlet Morning Glory

Habitat: Open tropical areas

Uses: Ornamental, Medicinal

IUCN Threat listing: Not Assessed (NA)

Description: Annual twining herb, stem up to 5 m; Leaves long-petioled, lamina ovate to circular in outline, base cordate, apex acuminate; Inflorescences lateral, sepals with almost equal awn, corolla salver-shaped, scarlet, stamens and style exserted; Capsules spherical, seeds densely pubescent, black.

Flowering and Fruiting: September-January

219. Scientific Name: *Ipomoea nil* (L.) Roth

Synonym: *Convolvulus nil* L.

Family: Convolvulaceae

Common Name: Blue Morning Glory

Habitat: Open tropical areas

Uses: Medicinal

IUCN Threat listing: Not Assessed (NA)

Description: Twining or spreading shrub; Stems retrosely hairy; Leaves ovate-cordate, 3-lobed; Flowers deep blue, tinged with pink, on 1 to 5-flowered peduncles; Capsules 3-celled, sub-globose or ovoid, glabrous.

Flowering and Fruiting: November-January

220. Scientific Name: *Ipomoea pes-tigridis* L.

Synonym: *Convolvulus pes-tigridis* (L.) Spreng.

Family: Convolvulaceae

Common Name: Tigerfoot Morning Glory

Habitat: Open tropical areas

Uses: Medicinal

IUCN Threat listing: Not Assessed (NA)

Description: Spreading or twining annual; Stem hairy/hispid Leaves deeply palmate, 5-9-lobed; Flowers sessile, white or pinkish, 3-many in peduncled heads, surrounded by large, hairy bracts; Capsules hidden in the calyx.

Flowering and Fruiting: September-December

221. Scientific Name: *Ixora chinensis* Lam.

Synonym: *Pavetta chinensis* (Lam.) Roem. & Schult.

Family: Rubiaceae

Common Name: Jungle Geranium

Habitat: Open tropical areas

Uses: Ornamental, Medicinal

IUCN Threat listing: Not Assessed (NA)

Description: Evergreen shrub, profusely branched; Leaves opposite, glossy, leathery, oblong, margins entire; Flowers produced almost all year long in dense, globose clusters, scarlet, tube long.

Flowering and Fruiting: All year round

222. Scientific Name: *Jacaranda mimosifolia* D. Don

Synonym: *Jacaranda chelonia* Griseb.

Family: Bignoniaceae

Common Name: Blue Jacaranda

Habitat: Humid tropical areas

Uses: Avenue tree, Fuel wood, Timber, Medicinal

IUCN Threat listing: Vulnerable

Description: Deciduous or evergreen tree, 5-15 m tall, bark thin, grey-brown, smooth at first, eventually becomes finely scaly; Leaves bi-pinnately compound leaves, up to 45 cm long; Flowers in dense terminal clusters, making the entire tree look lavender blue, tubular, 2.5 cm long; Fruit a round, flat, reddish brown, woody capsule, containing numerous small winged seeds.

Flowering and Fruiting: September-January

223. Scientific Name: *Jasminum sambac* (L.) Aiton

Synonym: *Mogorium sambac* (L.) Lam.

Family: Oleaceae

Common Name: Jasmine

Habitat: Open tropical areas

Uses: Aromatic, Medicinal

IUCN Threat listing: Not Assessed (NA)

Description: A scrambling, evergreen shrub, growing up to 3 m, stem glossy, dark green, twining; Leaves shiny, dark green, pointed, oval, opposite or whorled in 3; Flowers intensely fragrant, white, waxy, appearing in clusters; Fruits small, black berries.

Flowering and Fruiting: March-August

224. Scientific Name: *Jatropha curcas* L.

Synonym: *Manihot curcas* (L.) Crantz

Family: Euphorbiaceae

Common Name: Physic Nut

Vernacular Name: **Ratanjot**

Habitat: Open tropical areas

Uses: Hedge species, Oil

IUCN Threat listing: Not Assessed (NA)

Description: A perennial, poisonous shrub, up to 5 m high; Leaves large, broadly ovate, cordate, shallowly 3 lobed; Inflorescence axillary, flowers greenish yellow, female flowers usually slightly larger; Capsules 3-4 cm, globose, 3-lobed, yellowish green.

Flowering and Fruiting: All year round

225. Scientific Name: *Jatropha gossypiifolia* L.

Synonym: *Adenoropium gossypiifolium* (L.) Pohl

Family: Euphorbiaceae

Common Name: Black Physic Nut

Habitat: Open tropical areas

Uses: Ornamental, Hedge species

IUCN Threat listing: Not Assessed (NA)

Description: An erect, woody, deciduous, perennial, stem sympodial, much branched laterally; Leaves alternate, green, bronze, red or bright purplish-red, petiolate, palmately 3-5 lobed; Flowers unisexual, pedicellate, terminal, in corymbose cymes; Fruit trilocular capsule.

Flowering and Fruiting: July-September

226. Scientific Name: *Justicia adhatoda* L.

Synonym: *Adhatoda vasica* Nees

Family: Acanthaceae

Common Name: Malabar Nut

Habitat: Open tropical areas

Uses: Edible, Medicinal, Hedge species

IUCN Threat listing: Not Assessed (NA)

Description: Dense, evergreen, profusely growing herb; Stem glabrous, terete, with many, ascending branches; Leaves elliptic/lanceolate, acute at both ends, dark green above, pale beneath, petioles stout; Flowers in dense, shortly-peduncled, bracteate spikes; Fruit a capsule, 2 cm or more in length, clavate, pubescent, 4-seeded.

Flowering and Fruiting: December-June

227. Scientific Name: *Justicia procumbens* L.

Synonym: *Ecbolium procumbens* L.

Family: Acanthaceae

Common Name: Water Willow

Habitat: Open tropical areas

Uses: Edible, Medicinal

IUCN Threat listing: Not Assessed (NA)

Description: A slender, often tufted, prostrate or ascending, branched perennial herb; Leaves elliptic to oblong-ovate, obtuse at both ends, entire or slightly crenate on margin; Flowers borne in terminal, rather dense, cylindric spikes, pink, corolla bifid.

Flowering and Fruiting: June-December

228. Scientific Name: *Kalanchoe blossfeldiana* v. Poelln.

Synonym: *Kalanchoe coccinea* (H. Perrier) P. Boiteau

Family: Crassulaceae

Common Name: Flaming Katy

Habitat: Open tropical areas

Uses: Ornamental, Wound healing, Medicinal

IUCN Threat listing: Not Assessed (NA)

Description: A bushy, evergreen, perennial succulent plant, 30-45 cm in height; Leaves opposite or sub-opposite, scallop-edged, ovate-oblong, simple; Flowers on small raised peduncles, variously coloured, tetramerous.

Flowering and Fruiting: January-July

229. Scientific Name: *Kalanchoe pinnata* (Lam.) Pers.

Synonym: *Crassula pinnata* L. *f.*

Family: Crassulaceae

Common Name: Bryophyllum

Habitat: Open tropical areas

Uses: Ornamental, Wound healing, Medicinal

IUCN Threat listing: Not Assessed (NA)

Description: A succulent, glabrous, glaucous perennial herb, 0.3-2 m high; Leaves opposite decussate and, only present on upper parts in mature plants, upper leaves often 3-5-foliate, lower ones simple; Flowers pendulous, calyx papery, much inflated, corolla yellowish-green, tinged with pink or reddish, twice as long as calyx; Fruits capsular, with 4 slender papery tubes, seeds ellipsoid-oblong, longitudinally striate and brown.

Flowering and Fruiting: January-March

230. Scientific Name: *Kyllingia brevifolia* Rottb.

Synonym: *Cyperus brevifolius* (Rottb.) Hassk.

Family: Cyperaceae

Common Name: Green Kyllinga

Habitat: Open tropical grasslands

Uses: Fodder, Medicinal

IUCN Threat listing: Not Assessed (NA)

Description: Annual herb; Stem short; Leaves longer than the culm, linear; Spike solitary, spikelets ovate-lanceolate, single-flowered; Nuts laterally compressed, obovate-ellipsoid.

Flowering and Fruiting: July-November

231. Scientific Name: *Lablab purpureus* (L.) Sweet

Synonym: *Dolichos lablab* L.

Family: Leguminosae

Common Name: Lablab Bean

Habitat: Open tropical areas

Uses: Edible, Medicinal

IUCN Threat listing: Not Assessed (NA)

Description: Twining, cultivated vine, usually runs as escape; Leaves trifoliate, leaflets purplish green, broad oval or triangular; Flowers arranged in loose clusters, pea-like, brilliant purple or white; Pods flat, curved, about 10 cm long, bright purple or green, seeds dark colored with conspicuous white eye.

Flowering and Fruiting: July-October

232. Scientific Name: *Lagascea mollis* Cav.

Synonym: *Nocca mollis* (Cav.) Jacq.

Family: Asteraceae

Common Name: Silk Leaf

Habitat: Open tropical areas

Uses: Ornamental, Medicinal

IUCN Threat listing: Not Assessed (NA)

Description: An annual herb, 50-100 cm tall; Leaves opposite, simple, ovate, 3-5 cm long, margin toothed, silky beneath; Flowers in capitula, heads 2-3 cm across, flowers tiny, white-cream colored, petals 5.

Flowering and Fruiting: August-March

233. Scientific Name: *Lagerstroemia indica* L.

Synonym: *Lagerstroemia chinensis* L.

Family: Lythraceae

Common Name: Indian Crape Myrtle

Habitat: Open tropical areas

Uses: Ornamental, Timber, Medicinal, Erosion control

IUCN Threat listing: Not Assessed (NA)

Description: A much branched, deciduous tree with a wide spreading, flat topped or rounded habit; Leaves small, smooth-edged, circular or oval, dark green, changing to yellow orange-red in autumn; Flowers borne in panicles, white, pink, mauve, purple or carmine with crimped petals; Capsules ellipsoidal, 4-6-valved.

Flowering and Fruiting: June-August

234. Scientific Name: *Lagerstroemia parviflora* Roxb.

Synonym: *Murtughas parviflora* Kuntze

Family: Lythraceae

Common Name: Small Flowered Crape Myrtle

Habitat: Open tropical areas

Uses: Fuel wood, Gum, Fiber

IUCN Threat listing: Not Assessed (NA)

Description: Deciduous trees, up to 15 m high, bark dark brown or greyish-black; Leaves simple, opposite, deciduous, petioles slender, glabrous, lamina obovate-elliptic or oblong; Flowers in lax axillary panicles, bisexual, white, fragrant; Fruit a capsule, 3-4 valved, ovoid, brown, seeds many.

Flowering and Fruiting: February-May

235. Scientific Name: *Lagerstroemia speciosa* (L.) Pers.

Synonym: *Lagerstroemia flos-reginae* Retz.

Family: Lythraceae

Common Name: Queen Crape Myrtle

Habitat: Open tropical areas

Uses: Ornamental, Timber, Medicinal, Erosion control

IUCN Threat listing: Not Assessed (NA)

Description: A fast-growing, medium-sized, deciduous tree up to 15 m; Leaves deciduous, oval to elliptic, apex acute; Flowers in erect panicles, 20–40 cm long, petals 6, white to purple.

Flowering and Fruiting: April-July

236. Scientific Name: *Lantana camara* L.

Synonym: *Camara vulgaris* Benth.

Family: Verbenaceae

Common Name: Lantana

Habitat: Open tropical areas

Uses: Edible, Handicraft, Ornamental, Medicinal

IUCN Threat listing: Not Assessed (NA)

Description: Straggling, aromatic shrub; Stem profusely branched, armed with small, recurved prickles on the branches; Leaves opposite, ovate-oblong, acute, serrate, scabrous above, sparsely pilose beneath, apex acute, base rounded; Flowers orange-coloured, in pedunculate, short, capitate spikes; Drupe greenish-blue, shining, 1-seeded.

Flowering and Fruiting: Throughout the year

237. Scientific Name: *Leea asiatica* (L.) Ridsdale

Synonym: *Phytolacca asiatica* L.

Family: Vitaceae

Common Name: Bandicoot Berry

Habitat: Humid tropical areas

Uses: Medicinal

IUCN Threat listing: Not Assessed (NA)

Description: An erect, gregarious shrub, with angular stem swollen above the nodes and internodes; Leaves pinnately compound, leaflets 3-5, ovate or ovate-oblong, serrate, tip sharp, base rounded or cordate; Flowers borne in short, cymes at the end of branches, greenish white, calyx united, cup-like, often glandular-tipped.

Flowering and Fruiting: March-November

238. Scientific Name: *Leea rubra* Bl.

Synonym: *Leea brunoniana* C. B. Cl.

Family: Vitaceae

Common Name: Red Leea

Habitat: Open tropical grasslands

Uses: Medicinal

IUCN Threat listing: Not Assessed (NA)

Description: A semi-woody shrub, up to 3 m tall; Leaves alternate, stalked, pinnate, 30-60 cm long, leaflets many; Flowering shoots compact, 4-16 cm long, flowers bright red, with a yellow central disc; Fruits subglobose berries, turning dark red or purple when ripe, 6-seeded.

Flowering and Fruiting: April-August

239. Scientific Name: *Leonotis nepetifolia* (L.) R. Br.

Synonym: *Phlomis nepetifolia* L.

Family: Lamiaceae

Common Name: Lion's Ear

Habitat: Open tropical areas

Uses: Ornamental, Edible, Medicinal

IUCN Threat listing: Not Assessed (NA)

Description: Erect, branched herb, up to 8 ft. tall; Leaves smooth, triangular, margins coarsely toothed margins; Flowers borne in rounded, stem circling, spiny clusters, orange, tubular flowers emerge out of the spiny heads.

Flowering and Fruiting: July-September

240. Scientific Name: *Leucaena leucocephala* (Lam.) de Wit

Synonym: *Mimosa leucocephala* Lam

Family: Leguminosae

Common Name: White Lead tree

Habitat: Humid tropical areas

Uses: Fuel wood, Edible, Fodder

IUCN Threat listing: Not Assessed (NA)

Description: Small tree; Stem branching from ground; Leaves bipinnate, pinnae 5-10 pairs, linear, leaflets 10-15 pairs, acute, membraneous, downy; Flowers in globose, clustered heads, yellowish-white; Pods straight, clustered, thin, flat, dark brown on ripening.

Flowering and Fruiting: Throughout the year

241. Scientific Name: *Leucas aspera* (Willd.) Link

Synonym: *Phlomis aspera* Willd.

Family: Lamiaceae

Common Name: Common Leucas

Habitat: Open tropical areas

Uses: Aromatic, Medicinal, Insecticide

IUCN Threat listing: Not Assessed (NA)

Description: Erect, hirsute herbs up to 40 cm tall; Leaves linear-lanceolate, puberulous, base cuneate, margin distantly serrate, apex acute; Flowers white, sessile or sub-sessile, in terminal and axillary whorls, corolla 1 cm long.

Flowering and Fruiting: November-February

242. Scientific Name: *Leucas cephalotes* (Roth) Spreng.

Synonym: *Phlomis cephalotes* Roth

Family: Lamiaceae

Common Name: Head Leucas

Habitat: Open tropical areas

Uses: Edible, Medicinal

IUCN Threat listing: Not Assessed (NA)

Description: Annual, erect, pubescent herb; Stems up to 60 cm high, much branched, angular; Leaves opposite, petiolate, elliptic-lanceolate, distantly crenate-serrate; Flowers white, in terminal, globose heads; Nutlets brownish, smooth.

Flowering and Fruiting: September-January

243. Scientific Name: *Limonia acidissima* L.

Synonym: *Feronia elephantum* Corrêa

Family: Rutaceae

Common Name: Elephant Apple

Habitat: Open tropical areas

Uses: Edible, Medicinal, Gum

IUCN Threat listing: Not Assessed (NA)

Description: Erect, slow-growing tree, with a few upward-reaching branches, bark ridged, fissured and scaly, sharp spines present on some of the zigzag twigs; Leaves deciduous, alternate, dark-green, leathery, often minutely toothed, blunt or notched at the apex, dotted with oil glands; Flowers borne in small, loose, terminal or lateral panicles, yellowish green, tinged with red; Fruit hard, woody, round to oval in shape.

Flowering and Fruiting: December-June

244. Scientific Name: *Lindernia dubia* (L.) Pennell

Synonym: *Gratiola dubia* L.

Family: Linderniaceae

Common Name: Yellow seed False Pimpernel

Habitat: Open tropical areas

Uses: Fodder

IUCN Threat listing: Least Concerned (LC)

Description: Annual, erect herb; Stem 4-sided; Leaves sessile, opposite, oblong, acute, serrtulose, glabrous; Flowers short-pedicelled, in terminal racemes, white; Capsule narrowly cylindrical, many seeded.

Flowering and Fruiting: July-December

245. Scientific Name: *Linum usitatissimum* L.

Synonym: *Linum arvense* Neck.

Family: Linaceae

Common Name: Flax

Habitat: Open tropical areas

Uses: Oil, Medicinal, Fiber

IUCN Threat listing: Not Assessed (NA)

Description: Annual herbaceous plant, about 1 m high; Leaves glaucous green, slender, lanceolate; Flowers pure pale blue, with five petals; Fruit round, dry capsule, containing several glossy brown seeds.

Flowering and Fruiting: June-September

246. Scientific Name: *Ludwigia perennis* L.

Synonym: *Jussiaea caryophyllaea* Lam.

Family: Onagraceae

Common Name: Perennial Water Primrose

Habitat: Humid tropical areas

Uses: Ornamental, Edible

IUCN Threat listing: Least Concerned (LC)

Description: Erect, perennial or annual, aquatic herb; Leaves opposite or spiral, simple, entire or toothed, stipules absent or reduced; Flowers 2-or 4-merous, bisexual, solitary or in axils of leaves, or in racemes, often subtended by reduced bracts, sepals 2-5(-7), valvate, petals 2-5(-7); Fruit a capsule, berry, or nut, loculicidal or indehiscent; seeds smooth or variosly papillose.

Flowering and Fruiting: Throughout the year

247. Scientific Name: *Madhuca longifolia* (Roxb.) A. Chev.

Synonym: *Madhuca indica* J.F. Gmel.

Family: Sapotaceae

Common Name: Indian Butter Tree

Vernacular Name: Mahua

Habitat: Open tropical areas

Uses: Oil, Medicinal, Beer

IUCN Threat listing: Not Assessed (NA)

Description: Large, deciduous tree; Bark with shallow, transverse wrinkles and vertical cracks, grey or blackish; Leaves long petiolate, oblanceolate, clustered at the ends of branches, short-acuminate, base cuneate; Flowers numerous at the end of branches, greenish white, drooping on woolly pedicels; Fruit a juicy berry, ovoid, green, 1-4-seeded.

Flowering and Fruiting: February-April

248. Scientific Name: *Magnolia champaca* (L.) Baill. *ex* Pierre

Synonym: *Michelia champaca* (L.) Baill. *ex* Pierre

Family: Magnoliaceae

Common Name: Champak

Vernacular Name: Champa

Habitat: Humid tropical areas

Uses: Avenue tree, Aromatic, Timber

IUCN Threat listing: Least Concerned (LC)

Description: A medium-sized tree, up to 30 m, bark grey, lenticellate; Leaves simple, alternate, spiral; petiole 1-3 cm long, stout, lamina elliptic-lanceolatate; Flowers solitary, axillary, large, yellow, fragrant; Follicles warty, 2-3 cm long, dehiscing dorsally, seed 1, scarlet.

Flowering and Fruiting: Throughout the year

249. Scientific Name: *Mallotus philippensis* (Lam.) Müll. Arg.

Synonym: *Croton philippensis* Lam.

Family: Euphorbiaceae

Common Name: Kamala Tree

Vernacular Name: Rori

Habitat: Open tropical areas

Uses: Fuel wood, Dye, Medicinal

IUCN Threat listing: Not Assessed (NA)

Description: Small, evergreen tree; Stem up to 7 m high with thin, rough, dark grey bark; Leaves alternate, simple, petiolate, ovate-elliptic, entire or toothed, acuminate, base cuneate-rounded; Flowers unisexual, in axillary racemes or panicles, brown or brick red, female without petals; Capsule ovoid, covered with a bright crimson powder.

Flowering and Fruiting: October-March

250. Scientific Name: *Malva sylvestris* L.

Synonym: *Malva acutiloba* C. Koch

Family: Malvaceae

Common Name: High Mallow

Habitat: Open tropical areas

Uses: Ornamental, Edible, Medicinal

IUCN Threat listing: Not Assessed (NA)

Description: A biennial or perennial plant, up to 1 m length; Leaves simple, alternate, edge of the leaf blade lobed, or lobed and dentate Flowers in axillary clusters, pentamerous, stamens more than 10, carpels many; Fruit a dry capsule, does not split open when ripe.

Flowering and Fruiting: July-September

251. Scientific Name: *Mangifera indica* L.

Synonym: *Mangifera amba* Forsk

Family: Anacardiaceae

Common Name: Mango

Vernacular Name: Aam

Habitat: Open tropical areas

Uses: Edible, Medicinal

IUCN Threat listing: Data deficient (DD)

Description: Large, evergreen tree; Stem with greyish bark and dense, spreading foliage; Leaves oblong-lanceolate, alternate, simple, entire, spathulate, dark green, with somewhat wavy lamina; Flowers in terminal panicles or fascicles, greenish-yellow, polygamous, small; Fruit a fleshy drupe, kidney-shaped, greenish, dark yellow or light red on ripening.

Flowering and Fruiting: March-July

252. Scientific Name: *Manilkara zapota* (L.) P. Royen

Synonym: *Achras zapota* L.

Family: Sapotaceae

Common Name: Sapodilla

Vernacular Name: Cheeku

Habitat: Open tropical areas

Uses: Edible, Medicinal

IUCN Threat listing: Not Assessed (NA)

Description: Evergreen tree with dense, widely spreading crown, 9-20 m tall; Leaves glossy, alternate, spirally clustered at the tips of the forked twigs, elliptic, pointed at both ends; Flowers small, bell-like, with 3 brown-hairy outer sepals and 3 inner sepals, enclosing the pale-green corolla and 6 stamens; Fruit nearly round, oval, ellipsoidal or conical.

Flowering and Fruiting: Throughout the year

253. Scientific Name: *Mansoa alliacea* (Desv. *ex* Beauv.) A. Gentry

Synonym: *Adenocalymma alliaceum* (Lam.) Miers

Family: Bignoniaceae

Common Name: Garlic Vine

Vernacular Name: Lehsun lata

Habitat: Open tropical areas

Uses: Aromatic, Edible, Antimicrobial, Medicinal

IUCN Threat listing: Not Assessed (NA)

Description: Ornamental evergreen climber, young stems quadrangular, mature cylindrical; Leaves opposite, 2-foliolate, sometimes with a terminal tendril, tri-fid, deciduous, leaflets elliptical or broadly elliptical, chartaceous, apex obtuse or obtusely acuminate; Flowers few, in axillary racemes, pedicellate, corolla violet-pink or lavender, infundibuliform; Capsule oblong, woody, light brown.

Flowering and Fruiting: Throughout the year

254. Scientific Name: *Martynia annua* L.

Synonym: *Carpoceras angulata* A. Rich

Family: Martyniaceae

Common Name: Devil's Claw

Habitat: Open tropical areas

Uses: Ornamental, Medicinal

IUCN Threat listing: Not Assessed (NA)

Description: Annual herb/undershrub; Stem up to 1 m high, viscid, densely glandular-hairy; Leaves large, long-petioled, ovate-orbicular, dentate, glandular-hairy, apex sub-obtuse, base cordate; Flowers rose-coloured, large, drooping, in short axillary and terminal racemes; Fruit ovoid, with two upturned horns.

Flowering and Fruiting: August-December

255. Scientific Name: *Mecardonia procumbens* (Mill.) Small

Synonym: *Bacopa chamaedryoides* (Kunth) Wettstein

Family: Plantaginaceae

Common Name: Baby Jump Up

Habitat: Open tropical areas

Uses: Fodder

IUCN Threat listing: Not Assessed (NA)

Description: A creeping perennial herb; Leaves opposite, ovate to elliptic, with a wedge-shaped base, and margin coarsely toothed; Flowers in leaf axils, petals 4, yellow with reddish lines entering the throat; Capsule oblong-cylindric, 2-valved.

Flowering and Fruiting: June-October

256. Scientific Name: *Melochia corchorifolia* L.

Synonym: *Lochemia corchorifolia* Arn.

Family: Malavaceae

Common Name: Chocolate Weed

Habitat: Open tropical areas

Uses: Edible, Fibre, Medicinal

IUCN Threat listing: Not Assessed (NA)

Description: Erect branched herb, young stem, leaf-stalks, bracts and sepals hairy; Leaves variable in size and shape, margin saw-toothed, pointed at tip; Flowers pink with a greenish-yellow center, petals 5, obovate, blunt or flat at tip; Capsule sub-spherical, 5- valved.

Flowering and Fruiting: July-April

257. Scientific Name: *Mentha piperita* L.

Synonym: *Mentha piperoides* Malinv.

Family: Lamiaceae

Common Name: Peppermint

Vernacular Name: Pudina

Habitat: Humid tropical areas

Uses: Aromatic, Edible, Medicinal

IUCN Threat listing: Not Assessed (NA)

Description: A herbaceous, rhizomatous perennial plant, 30–90 cm; Leaves dark green with reddish veins, apex acute, margin coarsely toothed; Flowers in whorls (verticellasters) around the stem, forming thick, blunt spikes, purple, corolla 4-lobed; Fruit carcerulus, with nutlets.

Flowering and Fruiting: June-August

258. Scientific Name: *Merremia emarginata* (Burm. *f.*) Hall. *f.*

Synonym: *Evolvulus emarginatus* Burm. *f.*

Family: Convolvulaceae

Common Name: Kidneyleaf Morning Glory

Habitat: Open tropical areas

Uses: Edible, Medicinal

IUCN Threat listing: Least Concerned (LC)

Description: A perennial, prostrate plant with slender stems; Leaves alternate, reniform to broadly ovate, cordate basally with a broadly rounded sinus and rounded basal lobes, somewhat emarginate apically; Flowers on very short peduncles, sepals obovate to orbicular, corolla yellow, campanulate; Fruits capsular, subglobose, 5-6 mm long.

Flowering and Fruiting: October-March

259. Scientific Name: *Mesosphaerum suaveolens* (L.) Kuntze

Synonym: *Hyptis suaveolens* (L.) Poit.

Family: Lamiaceae

Common Name: American Mint

Habitat: Open tropical areas

Uses: Aromatic, Insecticide

IUCN Threat listing: Not Assessed (NA)

Description: Perennial, suffrescent, villous shrub; Stems tetragonal, hispidly hairy; Leaves opposite, petioled, ovate, cordate, crenate, serrate; Flowers small, blue or purple, in axillary umbels or heads; Fruit of 4 dry nutlets.

Flowering and Fruiting: September-April

260. Scientific Name: *Mimosa pudica* L.

Synonym: *Mimosa hispidula* Kunth

Family: Leguminosae

Common Name: Touch Me Not

Vernacular Name: Chhui-mui

Habitat: Open tropical areas

Uses: Antimicrobial, Wound healing, Medicinal

IUCN Threat listing: Least Concerned (LC)

Description: Erect to prostrately growing herb forming tufts, stem slender, branching, and sparsely to densely prickly; Leaves bipinnately compound, pinnae 2-4, leaflets 10-26 per pinna; Flower heads pink or purple, arise from the leaf axils, globose to ovoid; Fruit a cluster of 2-8 pods, prickly on the margins.

Flowering and Fruiting: September-November

261. Scientific Name: *Mirabilis jalapa* L.

Synonym: *Jalapa congesta* Moench

Family: Nyctaginaceae

Common Name: Four O' Clock Plant

Habitat: Open tropical areas

Uses: Ornamental, Edible, Medicinal

IUCN Threat listing: Not Assessed (NA)

Description: Erect, annual herb, with tuberous roots; Stem much branched; Leaves petiolate, ovate-acuminate, lanceolate, entire, apex acuminate, base truncate-cordate; Flowers in cymes, purple, white or yellow or variously variegated; Fruit leathery, 1-seeded, wrinkled.

Flowering and Fruiting: Throughout the year

262. Scientific Name: *Mitragyna parvifolia* (Roxb.) Korth.

Synonym: *Stephegyne parvifolia* (Roxb.) Korth.

Family: Rubiaceae

Common Name: Kratom Tree

Vernacular Name: Mundi

Habitat: Open tropical areas

Uses: Medicinal, Furniture

IUCN Threat listing: Not Assessed (NA)

Description: Medium-sized or large, deciduous, sweet-scented tree; Trunk irregular, bark brownish or greenish-grey; Leaves variable in shape and size, oval, orbicular or obovate, coriaceous, glabrous or pubescent; Flowers in globose heads, white or light-yellow; Cocci 2, dehiscent.

Flowering and Fruiting: March-December

263. Scientific Name: *Moringa oleifera* Lam.

Synonym: *Guilandia moringa* L.

Family: Moringaceae

Common Name: Drumstick Tree

Vernacular Name: Munga

Habitat: Open tropical areas

Uses: Antimicrobial, Wound healing, Medicinal

IUCN Threat listing: Not Assessed (NA)

Description: Medium-sized tree; Stem with corky bark, soft wood, young parts tomentose; Leaves usually tri-pinnate, 30-60 cm long, petioles sheathing, pinnae and pinnulae opposite; Flowers white, honey-scented, in panicles; Capsules 20-60 cm long, pendulous, ribbed.

Flowering and Fruiting: July-October

264. Scientific Name: *Mucuna pruriens* (L.) DC.

Synonym: *Dolichos pruriens* L.

Family: Leguminosae

Common Name: Velvet Bean

Vernacular Name: Kewanch

Habitat: Open tropical areas

Uses: Mordant

IUCN Threat listing: Not Assessed (NA)

Description: A perennial herb, 25 cm tall, stem erect, entirely hairy; Leaves sessile, blade linear, margins crustaceous glandular; Flowers in lax spikes at branch ends; bracts longer than calyx; Calyx 5-ribbed; sepals nearly as long as tube, awn-shaped.

Flowering and Fruiting: October-January

265. Scientific Name: *Murdannia nudiflora* (L.) Brenan

Synonym: *Commelina nudiflora* L.

Family: Commelinaceae

Common Name: Dove weed

Habitat: Humid tropical areas

Uses: Fodder

IUCN Threat listing: Not Assessed (NA)

Description: Diffuse, nearly glabrous herb; Stems branched, rooting at nodes, glabrous or slightly pubescent; Leaves sessile, linear-lanceolate, apex sub-acute, base rounded, sheaths hairy on the margin, ciliate on the mouth; Flowers in cymose panicles; Capsules sub-globose, coriaceous.

Flowering and Fruiting: September-November

266. Scientific Name: *Murraya koenigii* (L.) Spreng.

Synonym: *Bergera koenigii* L.

Family: Rutaceae

Common Name: Curry Tree

Vernacular Name: Meethi Neem

Habitat: Open tropical areas

Uses: Aromatic, Edible, Medicinal

IUCN Threat listing: Not Assessed (NA)

Description: Large, deciduous shrub; Stem with thin dark-brown or greyish bark; Leaves pinnate, pale green with terete, pubescent rachis, leaflets ovate-lanceolate, base oblique, apex acuminate; Flowers white, sweet-scented, in terminal corymbose panicles or umbels; Fruit a sub-globose, rugose berry, shining, initially green, turning gradually to purplish-black on ripening.

Flowering and Fruiting: February-April

267. Scientific Name: *Murraya paniculata* (L.) Jacq.

Synonym: *Murraya exotica* L.

Family: Rutaceae

Common Name: Orange Jasmine

Vernacular Name: Hathil

Habitat: Open tropical areas

Uses: Aromatic, Hedge species, Medicinal

IUCN Threat listing: Not Assessed (NA)

Description: Small, tropical, evergreen shrub, up to 7 m tall; Leaves glabrous and glossy, leaflets 3-7, elliptic to cuneate-obovate to rhombic; Flowers few, terminal, corymbose, fragrant, petals recurved, white; Fruit fleshy, oblong-ovoid, red to orange.

Flowering and Fruiting: Throughout the year

268. Scientific Name: *Musa paradisiaca* L.

Synonym: *Musa balbisiana* var. *vittata* (W. Ackm. *ex* Rodigas) M.R. Almeida

Family: Musaceae

Common Name: French Plantain

Vernacular Name: Kela

Habitat: Humid tropical areas

Uses: Edible, Medicinal

IUCN Threat listing: Not Assessed (NA)

Description: Tall, monocarpic perennial herb, rhizome underground; Leaves simple, petiole long and strong, leaf blade is large and broad with sheathy leaf base; Inflorescence spadix, flowers protected by large, brightly coloured, spirally arranged, boat shaped bracts called spathes, sessile, trimerous, unisexual or bisexual; Fruit elongated fleshy berry, seeds not produced.

Flowering and Fruiting: Throughout the year

269. Scientific Name: *Mussaenda frondosa* L.

Synonym: *Gardenia frondosa* (L.) Lam.

Family: Rubiaceae

Common Name: Flag Bush

Habitat: Humid tropical areas

Uses: Ornamental, Edible, Medicinal

IUCN Threat listing: Not Assessed (NA)

Description: A small shrub; Leaves broadly ovate, base rounded, caudate acuminate at apex, sparsely hairy on both surfaces; Flowers 5-merous, bracts and bracteoles linear, hairy, sepals showy, white, pubescent, leaf-like structure.

Flowering and Fruiting: September-March

270. Scientific Name: *Nelumbo nucifera* Gaertn.

Synonym: *Nelumbium nelumbo* (L.) Druce

Family: Nelumbonaceae

Common Name: Lotus

Vernacular Name: Kamal

Habitat: Humid tropical areas

Uses: Ornamental, Edible, Medicinal, Phytoremediation

IUCN Threat listing: Not Assessed (NA)

Description: An emergent hydrophyte, rhizomes under water; Leaves long-petioled, blade circular peltate, medium green or blue-green, hairless, margins smooth, often undulating; Flower solitary, tepals about 15, pink, receptacle golden yellow, a dense ring of golden yellow stamens; Each flower forms a seedpod, turning dark brown at maturity.

Flowering and Fruiting: July-September

271. Scientific Name: *Neolamarckia cadamba* (Roxb.) Bosser

Synonym: *Anthocephalus cadamba* (Roxb.) Miq.

Family: Rubiaceae

Common Name: Common Bur Flower

Vernacular Name: Kadam

Habitat: Humid tropical areas

Uses: Ornamental, Timber, Aromatic

IUCN Threat listing: Not Assessed (NA)

Description: Deciduous tree, up to 20 m, bark pale brown, vertically grooved, exfoliating in rectangular flakes; Leaves opposite, decussate, lanceolate, cauducous, petiole stout, glabrous, lamina ovate or elliptic-oblong; Flowers bisexual, yellowish, in globose heads, calyx and corolla tubular; Fruit a membranous capsule, on a fleshy globose receptacle, orange yellow.

Flowering and Fruiting: June-September

272. Scientific Name: *Nerium indicum* Mill.

Synonym: *Nerium odoratum Lam.*

Family: Apocyanaceae

Common Name: Oleander

Vernacular Name: Kaner

Habitat: Open tropical areas

Uses: Ornamental, Medicinal

IUCN Threat listing: Not Assessed (NA)

Description: Evergreen shrub; Stem with silvery-grey bark; Leaves whorled, linear-lanceolate or oblong, shortly petioled, thickly coriaceous, midrib very prominent; Flowers red, white or rose-coloured, 3.5 cm in diameter, in large, terminal, racemose cymes; Fruit 14-16 cm long follicles, rigid.

Flowering and Fruiting: April-September

273. Scientific Name: *Nicandra physalodes* (L.) Gaertner

Synonym: *Physalodes physalodes* (L.) Britt.

Family: Solanaceae

Common Name: Apple of Peru

Habitat: Humid tropical areas

Uses: Medicinal

IUCN Threat listing: Not Assessed (NA)

Description: Annual, erect, smooth herb; Stem 30-90 cm high; Leaves petioled, ovate-lanceolate, coarsely and irregularly lobed and serrated; Flowers nearly 15 cm in diameter, bluish-purple, borne singly on axillary, recurved stalks; Berry globose, flattened, loosely enclosed by the enlarged, 5-angled calyx.

Flowering and Fruiting: July-October

274. Scientific Name: *Nyctanthes arbor-tristis* L.

Synonym: *Parilium arbor-tristis* (L.) Gaertn.

Family: Oleaceae

Common Name: Tree of Sadness

Vernacular Name: Parijat

Habitat: Open tropical areas

Uses: Avenue tree, Aromatic, Medicinal

IUCN Threat listing: Not Assessed (NA)

Description: Small deciduous tree; Stem much branched, twigs quadrangular, bark grey or greenish white; Leaves shortly-petiolate, ovate, opposite, acute, coriaceous, covered with whitish hairs; Flowers white, streaked with orange, fragrant, in peduncled heads arranged in trichotomous cymes; Capsules small, orbicular, compressed.

Flowering and Fruiting: All year round

275. Scientific Name: *Nymphaea lotus* L.

Synonym: *Nymphaea rubra* Roxb. *ex* Salisb.

Family: Nymphaeaceae

Common Name: White Egyptian Lotus

Habitat: Humid tropical areas

Uses: Edible, Medicinal

IUCN Threat listing: Not Assessed (NA)

Description: Perennial emergent hydrophyte, up to 45 cm high; Leaves round, dentate, with a notch at the petiole; Flowers slightly fragrant, 15-25 cm across, sepals 4, petals up to 20, white, anthers numerous, yellow; Fruit 4-6 cm in diameter, depressed-globose.

Flowering and Fruiting: June-August

276. Scientific Name: *Nymphoides indica* (L.) Kuntze

Synonym: *Menyanthes indica* L.

Family: Menyanthaceae

Common Name: Water Snow Flake

Habitat: Humid tropical areas

Uses: Edible, Medicinal

IUCN Threat listing: Least Concerned (LC)

Description: Perennial, fast growing hydrophyte; Leaves all cauline, fertile and floating, ovate-orbicular, base deeply cordate with narrow sinuses, fleshy, glossy green above, pale and gland-dotted beneath; Flowers bisexual, distylous, in umbellate clusters, corolla white with yellow throat; Capsules ellipsoid. Seeds discoid, brownish, tubercled.

Flowering and Fruiting: June-August

277. Scientific Name: *Ocimum basilicum* L.

Synonym: *Ocimum album* L.

Family: Lamiaceae

Common Name: Basil

Vernacular Name: Ram Tulsi

Habitat: Open tropical areas

Uses: Aromatic, Edible, Medicinal, Oil

IUCN Threat listing: Not Assessed (NA)

Description: Erect, perennial herb; Stem and foliage aromatic from oil glands, smooth or velvety; Leaves opposite, elliptic to lanceolate, entire or toothed, slightly hairy; Flowers purplish-white, in 6-10-flowered dichasial cymes at nodes; Nutlets 3-4, oblong with rounded ends, black.

Flowering and Fruiting: August-November

278. Scientific Name: *Ocimum tenuiflorum* L.

Synonym: *Ocimum sanctum* L.

Family: Lamiaceae

Common Name: Holy Basil

Vernacular Name: Tulsi

Habitat: Open tropical areas

Uses: Aromatic, Edible, Medicinal, Oil

IUCN Threat listing: Not Assessed (NA)

Description: Aromatic, woody herb; Stem much-branched, often with a purplish tinge; Leaves opposite, ovate, elliptic-oblong to oblong, minutely gland-dotted, hairy, entire or crenate-serrate; Flowers purplish-pink, whorled on nodes; Nutlets broad ellipsoid, smooth.

Flowering and Fruiting: September-March

279. Scientific Name: *Oligochaeta ramosa* (Roxb.) Jafri

Synonym: *Amberboa ramosa* (Roxb.) Jafri

Family: Asteraceae

Common Name: Spiny thistle

Vernacular Name: Nili Kantili

Habitat: Open tropical grasslands

Uses: Medicinal

IUCN Threat listing: Not Assessed (NA)

Description: Decumbent to suberect herb reaching upto 18-40 cm in height; leaves sessile, cuneate/attenuate at base, elliptic-oblong or oblanceolate; Flower are capitula 1.0-2.0 cm across, solitary, terminal; stalk 0.2-1.6 cm long, sulcate, scabrid, bracts 1-2, small; fruit are cypsels blackish-brown, obconic.

Flowering and Fruiting: October-March

280. Scientific Name: *Opuntia stricta* (Haw.) Haw.

Synonym: *Cactus dillenii* Ker Gawl.

Family: Cactaceae

Common Name: Erect Prickly Pear

Vernacular Name: Naagfani

Habitat: Open tropical areas

Uses: Ornamental, Hedge species, Edible, Medicinal

IUCN Threat listing: Least Concerned (LC)

Description: Upright or spreading, succulent shrub, 50-100 cm tall, stems much-branched, form a series of flattened cladodes; Leaves reduced to tiny, terete or cone-shaped structures, quickly shed from the developing cladodes; Flower up to 7 cm long, bright yellow, often with pink or red markings on the outer petals; Fruits green when immature, turn reddish-purple on maturity.

Flowering and Fruiting: August-January

281. Scientific Name: *Ottelia alismoides* (L.) Pers.

Synonym: *Stratiotes alismoides* L

Family: Hydrocharitaceae

Common Name: Duck Lettuce

Habitat: Humid tropical areas

Uses: Medicinal

IUCN Threat listing: Least Concerned (LC)

Description: Partly or fully submerged freshwater herb; Young leaves somewhat linear, becoming oblanceolate, ovate or broadly cordate with maturity, lower margins of leaves and edges of petioles often shallowly serrated to sharply toothed; Flowers wrapped within spathes composed of green bracts, petals white, pink, blue or purple, often tinged with yellow at the base; Fruits fleshy capsules, seeds numerous.

Flowering and Fruiting: November-April

282. Scientific Name: *Oxalis corniculata* L.

Synonym: *Oxalis albicans Kunth*

Family: Oxalidaceae

Common Name: Creeping Wood sorrel

Vernacular Name: Teen Paan

Habitat: Open tropical areas

Uses: Edible, Medicinal

IUCN Threat listing: Not Assessed (NA)

Description: Slender, pubescent, perennial herb; Stems long, creeping or sub-terranean with ascending shoots; Leaves trifoliate, leaflets sub-sessile, broader than long, base cuneate; Flowers solitary or in umbels; Fruit a capsule, linear-oblong, sub-cylindric, angular, bursting on touch.

Flowering and Fruiting: July-October

283. Scientific Name: *Oxalis latifolia* Kunth

Synonym: *Sassia latifolia* (Kunth) Holub

Family: Oxalidaceae

Common Name: Sorrel

Vernacular Name: **Khat Mithhi**

Habitat: Open tropical areas

Uses: Fodder, Medicinal

IUCN Threat listing: Not Assessed (NA)

Description: A perennial, bulbous or stoloniferous herb; Leaves arise on long petioles from ground level, each made up of three widely heart-shaped leaflets; Flowers few, violet, pedicellate, 3-5 together, in long-peduncled umbels.

Flowering and Fruiting: May-June

284. Scientific Name: *Panicum maximum* Jacq.

Synonym: *Urochloa maxima* (Jacq.) R.D.Webster

Family: Poaceae

Common Name: Guinea Grass

Habitat: Open tropical grasslands

Uses: Fodder

IUCN Threat listing: Not Assessed (NA)

Description: Densely clump-forming, perennial grass with erect or ascending culms, nodes usually densely hirsute; Leaves flat, bright green, glabrous on margins, sometimes hirsute on upper surface; Spikelets awnless, green or tinged with purple, with 2 florets of which only the upper is fertile.

Flowering and Fruiting: June-October

285. Scientific Name: *Parasopubia delphinifolia* (L.) H.P. Hofm. & Eb. Fisch.

Synonym: *Sopubia delphinifolia* (L.) G. Don

Family: Orobanchaceae

Common Name: Common Sopubia

Habitat: Open tropical grasslands

Uses: Ornamental

IUCN Threat listing: Not Assessed (NA)

Description: Erect, branched herb, up to 45 cm high; Leaves narrow, filiform, hirsute; Flowers pink or pinkish-white, axillary, pedicels slender, calyx tube ribbed; Capsule truncate at apex, seeds oblong, pitted.

Flowering and Fruiting: September-November

286. Scientific Name: *Parthenium hysterophorus* L.

Synonym: *Parthenium pinnatifidum* Stokes

Family: Asteraceae

Common Name: Carrot Grass

Vernacular Name: Gajar Ghans

Habitat: Open tropical grasslands

Uses: Medicinal

IUCN Threat listing: Not Assessed (NA)

Description: A fast spreading, annual, pubescent herb, often a noxious weed; Stem profusely branched, up to 1 m high; Leaves alternate, deeply 1-2-pinnately dissected; Heads white, hairy, heterogamous, ray florets few; Achenes triquetrous, black, crowned, with remnant of corolla.

Flowering and Fruiting: October-March

287. Scientific Name: *Paspalidium flavidum* (Retz.) A.Camus

Synonym: *Panicum flavidum* Retz.

Family: Poaceae

Common Name: Yellow Watercrown Grass

Habitat: Open tropical grasslands

Uses: Fodder

IUCN Threat listing: Least Concerned (LC)

Description: Annual herb; Culms sub-erect, rooting below; Leaves linear, ciliated at the base, acuminate, sheaths longer than internodes, compressed, ligule a ring of hairs; Flowers in 6-9, sessile, distant, erect racemes, up to 2-2.5 cm long, arranged alternately; Lower florets male or barren, upper ones hermaphrodite.

Flowering and Fruiting: July-December

288. Scientific Name: *Paspalum distichum* L.

Synonym: *Digitaria disticha* (L.) Fiori & Paol.

Family: Poaceae

Common Name: Knot Grass

Habitat: Open tropical grasslands

Uses: Fodder, Erosion control

IUCN Threat listing: Least Concerned (LC)

Description: Fast-growing rhizomatous grass of wet areas, culms firm, sheaths loose, glabrous to pubescent; Leaf blades flat, tapering to an inrolled apex; The flowering stem bears terminal digitate racemes, up to 3 inches long, spikelets solitary, straw-purple in color, grain yellowish, oblong to ellipsoid.

Flowering and Fruiting: August-February

289. Scientific Name: *Paspalum scrobiculatum* L.

Synonym: *Paspalum amazonicum* Trin.

Family: Poaceae

Common Name: Kodo Millet

Vernacular Name: Kodo

Habitat: Open tropical grasslands

Uses: Edible, Medicinal

IUCN Threat listing: Least Concerned (LC)

Description: Tufted annual or perennial grass up to 150 cm tall, culms stout, erect, glabrous, somewhat bulbous at base, sheaths 7-14 cm long, glabrous or with sparse hairs at the collar, basal ones often purplish; Inflorescence of 4-6 racemes, 2-4 cm, alternate, distant, villous at base, fertile lemma indurated, finely pitted, caryopsis compressed-elliptic, pale.

Flowering and Fruiting: Throughout the year

290. Scientific Name: *Peltophorum pterocarpum* (DC.) K.Heyne

Synonym: *Peltophorum ferrugineum* (Decne.) Benth.

Family: Leguminosae

Common Name: Copper Pod

Vernacular Name: Pila Gulmohar

Habitat: Open tropical areas

Uses: Avenue tree, Antimicrobial, Medicinal, Dye

IUCN Threat listing: Not Assessed (NA)

Description: A deciduous tree, growing to 15-25 m tall; Leaves bipinnate, 30–60 cm long, pinnae 16-20, each pinna with 20-40 oval leaflets; Flowers yellow, in large compound racemes up to 20 cm long; Fruit a pod, red at first, ripening black, containing 1-4 seeds.

Flowering and Fruiting: March-September

291. Scientific Name: *Pennisetum hohenackeri* Hochst. *ex* Steud.

Synonym: *Cenchrus hohenackeri* (Hochst. ex Steud.) Morrone

Family: Poaceae

Common Name: Fountain Grass

Vernacular Name: Moa

Habitat: Open tropical grasslands

Uses: Handicrafts, Broom

IUCN Threat listing: Not Assessed (NA)

Description: A perennial herb, forming large bunches with thick tough roots; Leaf-blades narrowly linear, 15-40 cm long, rigid, tightly folded, ribbed on the outside; Flowers in linear panicles, 10-25 cm long, involucre encloses 1 stalkless, lance-shaped spikelet.

Flowering and Fruiting: March-December

292. Scientific Name: *Persicaria glabra* (Willd.) Gomez de la Maza

Synonym: *Polygonum glabrum* Willd.

Family: Polygonaceae

Common Name: Denseflower Knotweed

Habitat: Humid tropical areas

Uses: Medicinal

IUCN Threat listing: Least Concerned (LC)

Description: Erect, annual glabrescent herb; Stem soft and branched; Leaves petioled, lanceolate, sheathing, apex acuminate, narrowed towards base; Flowers pink or white, in terminal racemes; Nuts biconvex.

Flowering and Fruiting: July-September

293. Scientific Name: *Petalidium oblongifolium* C.B. Clakre

Synonym: *Petalidium oblongifolia*

Family: Acanthaceae

Common Name: Petal Bushes

Habitat: Open tropical areas

Uses: Ornamental

IUCN Threat listing: Not Assessed (NA)

Description: Small, erect herb, up to 1.5 m in length; Leaves borne on 4-10 mm long stalks, ovate, toothed, pointed to tapering; Flowers pale-blue or white, solitary, axillary or in small clusters; Capsule compressed, shortly stipitate, about 1.2 cm long, usually 2-seeded.

Flowering and Fruiting: April-August

294. Scientific Name: *Phoenix sylvestris* (L.) Roxb.

Synonym: *Elate sylvestris* L.

Family: Arecaceae

Common Name: Wild Date Palm

Vernacular Name: Chhindi

Habitat: Open tropical areas

Uses: Edible, Medicinal, Handicrafts

IUCN Threat listing: Not Assessed (NA)

Description: Tall tree, up to 15 m high; Stem bears persistent petiole bases and a large, thick, crown at top; Leaves alternate, pinnate, greyish-green, long, leaflets fascicled, rigid, spinous-pointed; Male flowers scented, white, spadix nearly 75 cm long, fruiting spadix longer, golden-orange in colour; Fruit oblong-ellipsoid, orange-yellow drupe, 1-seeded, edible.

Flowering and Fruiting: August-June

295. Scientific Name: *Phyllanthus emblica* L.

Synonym: *Emblica officinalis* Gaertn.

Family: Phyllanthaceae

Common Name: Gooseberry

Vernacular Name: Aanwla

Habitat: Open tropical areas

Uses: Edible, Medicinal, Mordant

IUCN Threat listing: Not Assessed (NA)

Description: Deciduous tree, up to 7 m high; Trunk crooked, with exfoliating, grey bark; Leaves small, sub-sessile, distichously arranged on deciduous branchlets appearing like pinnate leaves; Flowers greenish-yellow, unisexual, arranged in axillary fascicles; Fruits fleshy, globose, pale yellow.

Flowering and Fruiting: July-February

296. Scientific Name: *Phyllanthus fraternus* G.L.Webster

Synonym: *Phyllanthus lonphali* Benth., *nom. nud.*

Family: Phyllanthaceae

Common Name: Gulf Leaf Flower

Vernacular Name: Bhui Aanwla

Habitat: Open tropical areas

Uses: Medicinal, Dye

IUCN Threat listing: Not Assessed (NA)

Description: Glabrous, annual herb; Stems angular, branching from the base; Leaves sessile, distichous, overlapping, elliptic-oblong, glaucous beneath; Flowers axillary, unisexual, very minute, greenish-yellow, hanging down below the branchlets; Fruit a capsule, depressed, globose.

Flowering and Fruiting: August-December

297. Scientific Name: *Phyllanthus reticulatus* Poir.

Synonym: *Anisonema reticulatum* (Poir.) A.Juss.

Family: Phyllanthaceae

Common Name: Black Honey Shrub

Habitat: Humid tropical areas

Uses: Edible, Medicinal, Wound healing

IUCN Threat listing: Not Assessed (NA)

Description: Evergreen, much-branched somewhat climbing shrub; Leaves ovate-oblong to elliptic, produced on short lateral branchlets, looking like leaflets of a compound leaf; Flowers in clusters, on short axillary branchlets, small, yellowish, sexes separate on the same plant; Fruit berry-like, 4-6 mm across, blackish when ripe.

Flowering and Fruiting: August-December

298. Scientific Name: *Phyllodium pulchellum* (L.) Desv.

Synonym: *Desmodium pulchellum* (L.) Benth.

Family: Leguminosae

Common Name: Showy Desmodium

Habitat: Open tropical areas

Uses: Medicinal

IUCN Threat listing: Least Concerned (LC)

Description: Undershrubs or shrubs, up to 2 m tall; Leaves 3-foliolate, leaflets ovate, elliptic or ovate-oblong, base subacute, margins wavy, apex obtuse or subacute, glabrescent above; Racemes terminal or axillary, 7-16 cm long, bracts 10-18, leafy, orbicular, each enclosing 1-5 flowers, corolla yellowish; Pods indented on both sutures, sessile, slightly longer than broad, pubescent, seeds orbicular.

Flowering and Fruiting: July-November

299. Scientific Name: *Physalis lagascae* Roem. & Schult.

Synonym: *Physalis minima* L.

Family: Solanaceae

Common Name: Ground Cherry

Vernacular Name: Chirpoti

Habitat: Open tropical areas

Uses: Edible, Medicinal

IUCN Threat listing: Not Assessed (NA)

Description: An erect to prostrate, annual plant, can grow up to 100 cm; Leaves solitary or geminate, often slightly winged, rather sheathing at the base; Flowers solitary, on the side of the petiole, appearing axillary, erect to pendulous, corolla greenish-yellow or yellow, slightly purplish marked or apparently unmarked; Fruit greenish-yellow or pale yellow berry, sub-sessile, covered by inflated calyx.

Flowering and Fruiting: August-November

300. Scientific Name: *Piper longum* L.

Synonym: *Piper latifolium* Hunter

Family: Piperaceae

Common Name: Long Pepper

Vernacular Name: Pipli

Habitat: Humid tropical areas

Uses: Edible, Medicinal, Aromatic

IUCN Threat listing: Not Assessed (NA)

Description: Scandent or straggling shrubs, sometimes ascending and climbing; Leaves ovate, acute at apex, cordate, strongly oblique at base, 7-nerved, shining above; Female spike stout, erect, bracts peltate, orbicular, male spikes long, erect, slender; Berry 2 mm across, glabrous, black or deep red.

Flowering and Fruiting: July-October

301. Scientific Name: *Pithecellobium dulce* (Roxb.) Benth.

Synonym: *Inga dulcis* (Roxb.) Willd.

Family: Leguminosae

Common Name: Madras Thorn

Vernacular Name: Jangal Jalebi

Habitat: Open tropical areas

Uses: Edible, Medicinal

IUCN Threat listing: Not Assessed (NA)

Description: A small to medium-sized tree, up to 15-20 m in height, bark smooth, pale whitish-grey; Leaves bipinnate, pinnae 2, each with two leaflets; Flowers in small, dense, sub-spherical heads, 20-30 per head, pale whitish-green, stamen filaments white; Fruit spirally curved or coiled into 1-2 circles, bright red when ripe.

Flowering and Fruiting: December-February

302. Scientific Name: *Plectranthus barbatus* Andrews

Synonym: *Coleus barbatus* (Andrews) Benth. *ex* G.Don

Family: Lamiaceae

Common Name: Indian Coleus

Vernacular Name: Pathhar Chur

Habitat: Open tropical areas

Uses: Ornamental, Medicinal

IUCN Threat listing: Not Assessed (NA)

Description: An aromatic, erect herb; Leaves ovate, elliptic, apex obtuse, base cuneate, crenate, pubescent; Racemes terminal, bracts broadly ovate, acuminate, ciliate, corolla violet, lower lip broadly ovate, style bifid.

Flowering and Fruiting: October-December

303. Scientific Name: *Plectranthus mollis* (Aiton) Spreng.

Synonym: *Ocimum molle* Aiton

Family: Lamiaceae

Common Name: Soft Stem Mintleaf

Habitat: Open tropical areas

Uses: Antimicrobial, Medicinal

IUCN Threat listing: Not Assessed (NA)

Description: Erect, annual herb; Stem up to 70 cm high, velvety; Leaves opposite, petioled, broadly ovate, crenate-serrate, nearly cordate at base; Flowers pale lavender-blue, pedicelled, in cymes; Nutlets obovoid or ovate, 3-angled, orange-yellow.

Flowering and Fruiting: July-October

304. Scientific Name: *Plectranthus scutellarioides* (L.) R. Br.

Synonym: *Plectranthus blumei* (Benth.) Launert

Family: Lamiaceae

Common Name: Coleus

Habitat: Open tropical areas

Uses: Ornamental, Medicinal

IUCN Threat listing: Not Assessed (NA)

Description: An erect, perennial herb, up to 1.5 m; Leaves petiolate, variable in size, shape, and colour, usually brightly coloured and blotched, generally ovate-deltoid to broadly ovate; Flowers terminal racemes or panicles, corolla blue to purple or mauve, infundibular; Fruiting nutlets lenticular to broadly ovoid or subglobose, brown, glossy.

Flowering and Fruiting: August-October

305. Scientific Name: *Plumbago zeylanica* L.

Synonym: *Plumbago scandens* L.

Family: Plumbaginaceae

Common Name: Wild Leadwort

Vernacular Name: Chita Jadi

Habitat: Open tropical areas

Uses: Medicinal

IUCN Threat listing: Not Assessed (NA)

Description: Straggling, perennial undershrub; Stem up to 1 m high, usually needing support, striated; Leaves shortly-petioled, ovate-lanceolate, apex acute, base rounded; Flowers in long spike-like racemes, white, 1.5-2 cm long; Fruit a capsule, oblong, longitudinally furrowed, with a persistent calyx.

Flowering and Fruiting: November-March

306. Scientific Name: *Plumeria alba* L.

Synonym: *Plumeria revolutifolia* Stokes

Family: Apocyanaceae

Common Name: White Frangipani

Habitat: Open tropical areas

Uses: Ornamental, Medicinal, Oil

IUCN Threat listing: Not Assessed (NA)

Description: Somewhat succulent, evergreen shrub or a tree, branches soft and brittle, erect; Leaves strap-like, clustered at the end of branches; Flowers fragrant, white, with a small yellow center.

Flowering and Fruiting: November-July

307. Scientific Name: *Plumeria pudica* Jacq.

Synonym: *Plumeria cochleata* S.F.Blake

Family: Apocyanaceae

Common Name: Bridal Bouquet

Habitat: Open tropical areas

Uses: Ornamental

IUCN Threat listing: Not Assessed (NA)

Description: An erect, evergreen tree species reaching up to 6 m in height; Leaves fiddle-shaped with 2 lobes in the middle; Flowers white, with yellow throat, form a bouquet thus the common name 'Bridal Bouquet'.

Flowering and Fruiting: Throughout the year

308. Scientific Name: *Polianthes tuberosa* L.

Synonym: *Agave polianthes* Thiede & Eggli

Family: Asparagaceae

Common Name: Mexican Tuberose

Vernacular Name: Rajnigandha

Habitat: Open tropical areas

Uses: Ornamental, Edible, Oil

IUCN Threat listing: Not Assessed (NA)

Description: Perennial, erect, ornamental herb; Basal leaves long, bright green, clustered, upper leaves smaller, clasping the stem; Flowers in elongated spikes, clustered, fragrant, waxy white, bloom from bottom to top, tubular.

Flowering and Fruiting: August-December

309. Scientific Name: *Polyalthia longifolia* (Sonn.) Thwaites

Synonym: *Uvaria longifolia* Sonn.

Family: Annonaceae

Common Name: Mast Tree

Vernacular Name: Ashok

Habitat: Open tropical areas

Uses: Avenue tree, Fibre

IUCN Threat listing: Not Assessed (NA)

Description: An evergreen avenue tree, may attain a height more than 10 m; Leaves lanceolate, tapering to a fine point, margins undulate, glossy above, glabrous on both sides; Flowers in dense, hidden clusters, greenish yellow, petals narrow lanceolate; Fruit of 10-15 baccate monocarps on woody stalk.

Flowering and Fruiting: April-June

310. Scientific Name: *Polygala arvensis* Willd.

Synonym: *Polygala angustifolia* Buch.-Ham. *ex* Wall.

Family: Polygalaceae

Common Name: Field Milkwort

Habitat: Open tropical areas

Uses: Medicinal

IUCN Threat listing: Not Assessed (NA)

Description: Annual, small, stout, procumbent or decumbent herb; Stem weak, up to 25 cm long; Leaves round-oblong to linear, thick, leathery, shortly-stalked, apex mucronate, base sub-acute; Flowers in axillary racemes, pink, straight or drooping; Fruit a capsule, ovate, notched at apex, fringed, narrowly winged, 2-celled, enclosed in the calyx.

Flowering and Fruiting: Throughout the year

311. Scientific Name: *Polygonum plebejum* R. Br.

Synonym: *Polygonum plebeium* (Meisn.) M.R.Almeida

Family: Polygonaceae

Common Name: Common Knotweed

Habitat: Open tropical areas

Uses: Edible

IUCN Threat listing: Not Assessed (NA)

Description: Prostrate, annual, glabrous herb; Stem diffusely-branched; Leaves sessile or shortly-stalked, linear-elliptic, apex and base acute-acuminate; Flowers solitary, axillary, sessile, perianth pink; Nuts trigonous, shining.

Flowering and Fruiting: March-May

312. Scientific Name: *Pongamia pinnata* (L.) Pierre

Synonym: *Derris indica* (Lam.) Benn.

Family: Leguminosae

Common Name: Indian Beech

Vernacular Name: Karanj

Habitat: Open tropical areas

Uses: Fuel wood, Oil

IUCN Threat listing: Least Concerned (LC)

Description: Moderate-sized, glabrous, deciduos tree; Stem irregular with soft, smooth, grey bark; Leaves imparipinnate, bright green, glabrous, leaflets opposite, 5-7, ovate-oblong or elliptic, shortly acuminate; Flowers in axillary, pedunculate racemes; Pods indehiscent, thick, hard, 4-5 cm long, usually 1-seeded.

Flowering and Fruiting: April-June

313. Scientific Name: *Pontederia cordata* L.

Synonym: *Unisema cordata* (L.) Farw.

Family: Pontederiaceae

Common Name: Pickerel Weed

Habitat: Humid tropical areas

Uses: Ornamental, Edible, Medicinal

IUCN Threat listing: Not Assessed (NA)

Description: A perennial plant, up to 1 m, growing in marshy areas; Leaves basal, arising directly from the rootstock, blades of leaves highly variable; Flowers in a single spike of more than 100 flowers, blue-violet.

Flowering and Fruiting: May-August

314. Scientific Name: *Psidium guajava* L.

Synonym: *Myrtus guajava* (L.) Kuntze

Family: Myrtaceae

Common Name: Guava

Vernacular Name: Bihi

Habitat: Open tropical areas

Uses: Edible, Medicinal

IUCN Threat listing: Not Assessed (NA)

Description: Erect, woody shrub to a small tree; Stem with scaly, reddish-brown bark; Leaves opposite, coriaceaous, oblong or oblong-elliptic, shortly-petioled, veins prominent above; Flowers white, on axillary, 2-3-flowered peduncles; Fruits many-seeded, globose, ovoid or pyriform, yellow with white or pink pulp.

Flowering and Fruiting: March-May

315. Scientific Name: *Putranjiva roxburghii* Wall.

Synonym: *Drypetes roxburghii* (Wall.) Hurus.

Family: Putranjivaceae

Common Name: Officinal Drypetes

Vernacular Name: Jiyo Pota

Habitat: Open tropical areas

Uses: Avenue tree, Medicinal

IUCN Threat listing: Not Assessed (NA)

Description: Moderate-sized, evergreen tree; Stem with smooth, dark grey bark, drooping branches and slender, pubescent branchlets; Leaves shortly-petioled, alternate, bifarious, obliquely elliptic-oblong to ovate, coriaceous, acute, serrulate; Flowers greenish yellow, dioecious, male flowers in dense, axillary clusters, female only 1-3 in axils; Fruit a drupe, 1.2 cm long, ellipsoid, white-tomentose, hard.

Flowering and Fruiting: March-August

316. Scientific Name: *Pycreus polystachyos* (Rottb.) P. Beauv.

Synonym: *Cyperus polystachyos* Rottb.

Family: Cyperaceae

Common Name: Manyspike Flatsedge

Habitat: Humid tropical areas

Uses: Fodder

IUCN Threat listing: Least Concerned (LC)

Description: A rhizomatous, perennial or annual grass-like sedge, stems rigid; Leaves narrow, grass-like, often tufted at the base of the plant; Brown spikes at the tip of the stem have 3-6 green leafy bracts; The seed-head is an irregular cluster.

Flowering and Fruiting: December-January

317. Scientific Name: *Pyrus communis* L.

Synonym: *Pyrus sativa DC.*

Family: Rosaceae

Common Name: Pear

Vernacular Name: Naspati

Habitat: Open tropical areas

Uses: Edible, Medicene

IUCN Threat listing: Not Assessed (NA)

Description: Erect, medium-sized to large tree; Stem with grey, fissured bark, branched from base; Leaves dark green, broad-ovate to elliptic, long-petioled, acuminate; Flowers white, in clusters; Fruit a sub-globose pome, yellow, with remnants of calyx on the tip.

Flowering and Fruiting: April-August

318. Scientific Name: *Radermachera xylocarpa* (Roxb.) Roxb. *ex* K. Schum.

Synonym: *Bignonia xylocarpa* Roxb.

Family: Bignoniaceae

Common Name: Padri Tree

Vernacular Name: Garur Phal

Habitat: Open tropical areas

Uses: Timber, Medicinal

IUCN Threat listing: Not Assessed (NA)

Description: A large deciduous tree, up to 10 m tall, stem short and erect, with gray, soft, scaly bark; Leaves opposite, double-pinnate, leaflets 3-9, elliptic, oblong, sharp tipped, with a rounded base; Flowers in dense panicles, at the end of branches, fragrant white, funnel shaped, white, tinged with yellow; Capsules are long, up to 1 m, 5 cm wide, curved and woody.

Flowering and Fruiting: March-May

319. Scientific Name: *Rauvolfia serpentina* (L.) Benth. *ex* Kurz

Synonym: *Ophioxylon serpentinum* L.

Family: Apocyanaceae

Common Name: Indian Snakeroot

Vernacular Name: Sarpagandha

Habitat: Open tropical areas

Uses: Ornamental, Medicinal

IUCN Threat listing: Not Assessed (NA)

Description: An erect, annual shrub, up to 60 cm tall; Leaves in whorls of 3, thin, lanceolate, acute, bright green above and pale beneath; Flowers in irregular corymbose cymes, white, often tinged with violet; Fruits drupaceous, single or clustered, shining black.

Flowering and Fruiting: March-December

320. Scientific Name: *Rhaphidophora decursiva* (Roxb.) Schott

Synonym: *Monstera decursiva* (Roxb.) Schott

Family: Araceae

Common Name: Creeping Philodendron

Habitat: Humid tropical areas

Uses: Ornamental

IUCN Threat listing: Not Assessed (NA)

Description: A creeping liana, up to 20 m long, pseudostem about 20 cm long, green, mottled grey and pink; Leaf solitary, exceptionally bifoliate, cataphylls purple, leaf blade radiate, leaflets 7–11, ovate, margins entire, apex acute, acumen 1 cm; Inflorescence held below the foliage, green and pink mottled, spathe tube cylindrical, spathe mouth margins recurved; Fruits red-orange when ripe.

Flowering and Fruiting: July-October

321. Scientific Name: *Ricinus communis* L.

Synonym: *Croton spinosus* L.

Family: Euphorbiaceae

Common Name: Castor Plant

Vernacular Name: Arandi

Habitat: Open tropical areas

Uses: Oil, Medicinal

IUCN Threat listing: Not Assessed (NA)

Description: Evergreen shrub; Stem hollow; Leaves alternate, long-petiolate, glabrous, peltate, palmately lobed, lobes irregularly serrate; Inflorescence a terminal panicle, with upper female and lower male flowers; Capsule sub-globose, echinate, seeds oblong, with a white caruncle.

Flowering and Fruiting: September-December

322. Scientific Name: *Rosa indica* L.

Synonym: *Rosa bengalensis* Pers.

Family: Rosaceae

Common Name: Rose Plant

Vernacular Name: Gulab

Habitat: Open tropical areas

Uses: Ornamental, Aromatic, Medicinal

IUCN Threat listing: Not Assessed (NA)

Description: Annual or perennial, prostrate, scandent, climbing or erect shrub; Leaves alternate, rarely opposite, simple, sometimes pinnately compound, stipulate, stipule may be minute and caducous; Flowers pedicillate, bracteate, actinomorphic; Fruit fleshy, red and edible.

Flowering and Fruiting: Throughout the year

323. Scientific Name: *Rotheca serrata* (L.) Steane & Mabb.

Synonym: Clerodendrum serratum (L.) Moon

Family: Lamiaceae

Common Name: Blue Fountain Bush

Vernacular Name: Bharangi

Habitat: Open tropical areas

Uses: Edible, Medicinal

IUCN Threat listing: Not Assessed (NA)

Description: A small bush, up to 8 feet; Leaves in opposite pairs, or 3 at a node, oval, serrated, hairless, base acute, petiole stout; Flowers showy, in cymes, covered with short, soft, erect hairs, corolla blue, cylindrical, hairless outside, hairy inside, stamens arching out; Fruit a 4-lobed fleshy drupe, green at first and black when ripe.

Flowering and Fruiting: August-January

324. Scientific Name: *Rubia cordifolia* L.

Synonym: *Galium cordifolium* (L.) Kuntze

Family: Rubiaceae

Common Name: Indian Madder

Vernacular Name: Manjistha

Habitat: Humid tropical areas

Uses: Dye, Medicinal

IUCN Threat listing: Not Assessed (NA)

Description: A creeping herb, may attain a length of 1.5 m; Leaves evergreen, in whorls of 4-7, star-like around the central stem; Flowers in dense racemes, small, petals 5, pale yellow petals; Fruits red to black berries, 4-6 mm in diameter.

Flowering and Fruiting: August-November

325. Scientific Name: *Ruellia tuberosa* L.

Synonym: *Ruellia clandestina* L.

Family: Acanthaceae

Common Name: Minnie Root

Habitat: Humid tropical areas

Uses: Ornamental, Medicinal

IUCN Threat listing: Not Assessed (NA)

Description: Tender, perennial, suffruticose herb up to 90 cm tall; Petiole glabrous, blade ovate-triangular, serrate, glandular above, base truncate, apex acuminate; Racemes 2-6-flowered, up to 20 cm, bracts ovate, of the same colour as corolla, petals pubescent, tube slightly dilated at throat, style gynnobasic.

Flowering and Fruiting: July-October

326. Scientific Name: *Rumex dentatus* L.

Synonym: *Rumex nipponicus* Franch. & Sav.

Family: Polygonaceae

Common Name: Toothed Dock

Habitat: Humid tropical areas

Uses: Edible, Medicinal

IUCN Threat listing: Not Assessed (NA)

Description: Erect annual herb; Stem grooved, glabrous, usually tinged with red; Leaves petiolate, base sheathing, cordate-truncate, apex sub-acute; Flowers in whorls, shortly-pedicelled, bisexual, perianth broadly ovate; Nuts trigonous.

Flowering and Fruiting: January-May

327. Scientific Name: *Saccharum spontaneum* L.

Synonym: *Imperata spontanea* (L.) P. Beauv.

Family: Poaceae

Common Name: Wild Sugarcance

Vernacular Name: Kans

Habitat: Open tropical areas

Uses: Medicinal

IUCN Threat listing: Least Concerned (LC)

Description: Strongly rhizomatous, tufted perennial; Culms up to 3 m high, solid, glabrous, hard, pithy; Leaves linear-lanceolate, glaucous beneath, scarbid on midrib, sheaths auricled, ligule membraneous; Inflorescence a silvery-silky panicle, up to 70 cm long; Spikelets paired, awnless, one sessile, the other pedicelled, 2-flowered.

Flowering and Fruiting: November-February

328. Scientific Name: *Salvia splendens* Sellow *ex* Schult.

Synonym: *Salvia brasiliensis* Spreng.

Family: Lamiaceae

Common Name: Scarlet Sage

Habitat: Open tropical areas

Uses: Ornamental, Medicinal

IUCN Threat listing: Not Assessed (NA)

Description: Tender, perennial, suffruticose herb up to 90 cm tall; Petiole glabrous, blade ovate-triangular, serrate, glandular above, base truncate, apex acuminate; Racemes 2-6-flowered, up to 20 cm, bracts ovate, of the same colour as corolla, petals pubescent, tube slightly dilated at throat, style gynnobasic.

Flowering and Fruiting: January-September

329. Scientific Name: *Sapindus mukorossi* Gaertn.

Synonym: *Sapindus detergens* Roxb.

Family: Sapindaceae

Common Name: North Indian Soapnut

Vernacular Name: Reetha

Habitat: Open tropical areas

Uses: Cosmetic, Insecticidal, Medicinal, Surfactant

IUCN Threat listing: Not Assessed (NA)

Description: Deciduous tree; Stem branched, up to 25 m high; Leaves long stalked, odd pinnate, leaflets 5-10 pairs; Flowers small, greenish white, polygamous, in terminal panicles; Fruits solitary, round nuts, 2-2.5 cm diameter, fleshy, yellowish brown in color.

Flowering and Fruiting: October-January

330. Scientific Name: *Saraca indica* L.

Synonym: *Jonesia pinnata* Willd.

Family: Leguminosae

Common Name: Sorrowless Tree

Vernacular Name: Ashok

Habitat: Humid tropical areas

Uses: Ornamental, Edible, Medicinal

IUCN Threat listing: Not Assessed (NA)

Description: Small tree with compact crown, up to 10 m high; Leaves pinnately compound, leaflets 6-12, reddish when young; Flowers, fragrant, in big dense clusters directly on trunk or branches, petals absent, sepals open yellow, aging to orange and red; Fruit large, flat, leathery pods, 20 cm, purplish when mature.

Flowering and Fruiting: January-May

331. Scientific Name: *Scoparia dulcis* L.

Synonym: *Gratiola micrantha* Nuttall

Family: Plantaginaceae

Common Name: Sweet Broom Weed

Vernacular Name: Mithi Patti

Habitat: Open tropical areas

Uses: Medicinal

IUCN Threat listing: Not Assessed (NA)

Description: Erect perennial herb of moist and open areas; Leaves decussate to whorled, obovate-oblong, base attenuate, apex acute, margin crenate-serrate; Flowers axillary, solitary or 2, pedicels slender, calyx lobes 2-3 mm, ovate-oblong, ciliate along margin, petals white, 4-6 mm long, rotate, throat densely hairy; Capsules 1-2 mm across, globose.

Flowering and Fruiting: Throughout the year

332. Scientific Name: *Semecarpus anacardium* L. *f.*

Synonym: *Anacardium latifolium* Lam.

Family: Anacardiaceae

Common Name: Marking Nut

Vernacular Name: Bhelva

Habitat: Open tropical areas

Uses: Edible, Medicinal

IUCN Threat listing: Not Assessed (NA)

Description: Medium to large sized deciduous tree; Leaves simple, alternate, oblong, rounded at apex, glabrous above and pubescent below; Flower greenish white, in panicles; Fruits smooth, ovoid, lustrous black, about 2.5 cm long.

Flowering and Fruiting: February-April

333. Scientific Name: *Senegalia pennata* (L.) Maslin

Synonym: *Acacia pennata* (L.) Willd.

Family: Leguminosae

Common Name: Climbing Acacia

Vernacular Name: Biswal

Habitat: Open tropical areas

Uses: Fodder, Edible

IUCN Threat listing: Not Assessed (NA)

Description: Perennial climber; Stem twining, prickly, prickles slightly recurved, dark and shining; Pinnae 8-15 pairs, leaflets 40-50 pairs, sessile, linear; Flowers in globose, white or pale yellow heads; Pods 10-15 cm long, linear-oblong, acute at both ends, flat, thin, glabrous.

Flowering and Fruiting: October-January

334. Scientific Name: *Senna alata* (L.) Roxb.

Synonym: *Cassia alata* L.

Family: Leguminosae

Common Name: Candle Bush

Vernacular Name: Dad Murdan

Habitat: Humid tropical areas

Uses: Medicinal, Insecticidal

IUCN Threat listing: Not Assessed (NA)

Description: A deciduous shrub, up to 6 m tall; Leaves pinnately compound, 50–80 cm, dark green; Flowers in erect axillary racemes, bright yellow, 5-petaled, cup-shaped; Fruits almost straight pods, up to 25 cm long.

Flowering and Fruiting: September-January

335. Scientific Name: *Senna auriculata* (L.) Roxb.

Synonym: *Cassia auriculata* L.

Family: Leguminosae

Common Name: Tanner's Cassia

Vernacular Name: Tarwar

Habitat: Open tropical areas

Uses: Ornamental, Medicinal, Dye

IUCN Threat listing: Not Assessed (NA)

Description: Evergreen shrub, ornamental; Leaves alternate, stipulate, paripinnately compound, numerous, closely placed, rachis narrowly furrowed, slender, pubescent; Flowers irregular, bisexual, bright yellow and large, pedicels glabrous; Fruit a short legume, 7.5–11 cm long, 1.5 cm broad, oblong, obtuse.

Flowering and Fruiting: Throughout the year

336. Scientific Name: *Senna occidentalis* (L.) Link

Synonym: *Cassia occidentalis* (L.) Rose

Family: Leguminosae

Common Name: Coffee Senna

Vernacular Name: Kasunda

Habitat: Open tropical areas

Uses: Edible, Fodder, Medicine

IUCN Threat listing: Not Assessed (NA)

Description: Diffuse undershrub; Stem up to 1 m high, with purplish branches; Leaves paripinnate, leaflets 6-10, opposite, shortly stalked, ovate-oblong, apex acuminate, base slightly oblique; Flowers in axillary corymbs, yellow, faintly veined with orange; Pod flattened, recurved, 10-12 cm long, compressed, transversely septate.

Flowering and Fruiting: August-December

337. Scientific Name: *Senna siamea* (Lam.) H.S. Irwin & Barneby

Synonym: *Cassia siamea* Lam.

Family: Leguminosae

Common Name: Siamese Cassia

Vernacular Name: Kasod

Habitat: Open tropical areas

Uses: Avenue tree, Medicinal

IUCN Threat listing: Not Assessed (NA)

Description: Moderate sized tree; Stem with grey bark; Leaves paripinnate, eglandular, leaflets oblong, glaucous beneath, glabrous; Flowers with bright yellow petals, bearing faint red lines on the interior, obovate, rounded; Pods flat, nearly straight, many seeded, with thickened sutures.

Flowering and Fruiting: March-September

338. Scientific Name: *Senna tora* (L.) Roxb.

Synonym: *Cassia tora* L.

Family: Leguminosae

Common Name: Sickle Senna

Vernacular Name: Chikoda

Habitat: Open tropical areas

Uses: Edible, Oil, Medicinal

IUCN Threat listing: Not Assessed (NA)

Description: Annual, glaucous herb; Stem branched; Leaves with grooved rachis, caduceus stipules, leaflets opposite, obovate, rounded; Flowers solitary, axillary or more commonly in pairs; Pods nearly tetragonous, puberulous, not reticulate, 25-30-seeded.

Flowering and Fruiting: August-December

339. Scientific Name: *Sesamum indicum* L.

Synonym: *Sesamum orientale* L.

Family: Pedaliaceae

Common Name: Sesame

Vernacular Name: Til

Habitat: Open tropical areas

Uses: Edible, Oil, Medicinal

IUCN Threat listing: Not Assessed (NA)

Description: Annual, erect, glandular-pubescent herb; Stem 30-60 cm high, glandular; Leaves opposite, usually deeply lobed below, alternate, linear-oblong above, lobed, glabrous; Flowers 1-3 together in leaf axils or in racemes, panicled, purple or pinkish; Capsule oblong, valvular, glandular-pubescent, erect, shortly-beaked.

Flowering and Fruiting: August-October

340. Scientific Name: *Sesbania sesban* (L.) Merr.

Synonym: *Aeschynomene sesban* L.

Family: Leguminosae

Common Name: Common Sesban

Vernacular Name: Jayanti

Habitat: Open tropical areas

Uses: Edible, Medicinal, Fibre

IUCN Threat listing: Not Assessed (NA)

Description: A small, often multi-stemmed shrub or tree, growing to 4-8 m high; Leaves pinnately compound, 2-18 cm, leaflets linear oblong, 6-27 pairs; Flowers in 2-20 flowered racemes, yellow with purple or brown streaks on the corolla; Pods sub-cylindrical, straight or slightly curved, up to 30 cm long, with 10-50 seeds.

Flowering and Fruiting: July-November

341. Scientific Name: *Setaria helvola* (L. *f.*) Roem. & Schult.

Synonym: *Setaria pumila* (Poir.) Roem. & Schult

Family: Poaceae

Common Name: Yellow Foxtail

Vernacular Name: Ankutt

Habitat: Open tropical grasslands

Uses: Fodder

IUCN Threat listing: Not Assessed (NA)

Description: Annual grass, culms 30-100 cm high; Leaves linear or lanceolate, base rounded, apex acuminate, sparsely hairy,sheaths keeled;

Spikelets 2-2.5 mm long, ovate or subglobose; Lower glume ovate-acute, upper ovate or orbicular, lower floret male or barren, upper floret bisexual.

Flowering and Fruiting: July-October

342. Scientific Name: *Shorea robusta* Gaertn.

Synonym: *Dryobalanops robusta* (Gaertn. *f.*) Oken

Family: Dipterocarpaceae

Common Name: Sal Tree

Vernacular Name: Sarai

Habitat: Open tropical areas

Uses: Timber, Furniture, Oil, Medicinal

IUCN Threat listing: Least Concerned (LC)

Description: Huge, deciduous tree; Trunk branched high up, bark smooth when young, rough of old trees; Leaves alternate, ovate-oblong, acuminate, tough, thinly coriaceous, glabrous and shiny, base cordate; Flowers yellowish, sub-sessile, in lax terminal or axillary panicles; Fruit ovoid, acute, fleshy, indehiscent, seeds winged.

Flowering and Fruiting: March-April

343. Scientific Name: *Sida acuta* Burm. *f.*

Synonym: *Sida orientalis* DC.

Family: Malvaceae

Common Name: Wireweed

Vernacular Name: Kharheta

Habitat: Open tropical areas

Uses: Broom, Medicinal, Fibre

IUCN Threat listing: Not Assessed (NA)

Description: Small, erect undershrub; Stem much branched, about 60-90 cm high, branches smooth or slightly rough, stellately hairy; Leaves lanceolate-oblong, rounded at the base, acute or obtuse, crenate-serrate, glabrous; Flowers solitary, axillary or in pairs, orange-yellow, calyx glabrous; Schizocarps with less than 10 mericarps, rugose.

Flowering and Fruiting: September-December

344. Scientific Name: *Sida cordifolia* L.

Synonym: *Malvinda cordifolia* (L.) Medic.

Family: Malvaceae

Common Name: Heart Leaf Sida

Vernacular Name: Barial

Habitat: Open tropical areas

Uses: Medicinal

IUCN Threat listing: Not Assessed (NA)

Description: A perennial, robust herb/erect undershrub; Stem much branched from the base, stellate-tomentose; Leaves 2.5-5.0 cm long, ovate, cordate, crenate-serrate, stellately hairy on both surfaces; Flowers small, pale yellow or nearly white, solitary or in axillary fascicles; Fruit a schizocarpic capsule, mericarps 8-10, ripe carpels each with 2 awns.

Flowering and Fruiting: October-December

345. Scientific Name: *Sida rhombifolia* L.

Synonym: *Malva rhombifolia (L.) E.H.L. Krause*

Family: Malvaceae

Common Name: Jelly Leaf

Vernacular Name: Sehdev

Habitat: Open tropical areas

Uses: Broom, Medicinal, Fibre

IUCN Threat listing: Not Assessed (NA)

Description: Small, erect undershrub; Stem much branched, stellately hairy, branches ascending; Leaves variable, rhomboid or obovate, glabrous above, tomentose beneath, base cuneate, margins dentate; Flowers solitary, axillary, light yellow in colour; Schizocarps sub-globose, mericarps 7-10 with 1-2 awns.

Flowering and Fruiting: September-December

346. Scientific Name: *Smithia conferta* Sm.

Synonym: *Smithia geminiflora* Roth

Family: Leguminosae

Common Name: Paired Flower Smithia

Vernacular Name: Laxman Buti

Habitat: Humid tropical areas

Uses: Fodder, Medicinal

IUCN Threat listing: Not Assessed (NA)

Description: Annual herb, up to 3-4 ft long; Branches long, straight, rod-like, straggling, smooth, lower internodes long; Leaves small, alternate, compound, leaflets 8-16, oblong, blunt, with shortly pointed tip; Flowers yellow, in axillary pairs on upper leaves, crowded at the end of branches.

Flowering and Fruiting: April-August

347. Scientific Name: *Solanum nigrum* L.

Synonym: *Solanum humile* Salisb.

Family: Solanaceae

Common Name: Black Nightshade

Vernacular Name: Mukkaiyan

Habitat: Open tropical areas

Uses: Edible, Medicinal

IUCN Threat listing: Not Assessed (NA)

Description: Erect, glabrous, much branched, unarmed herb; Stem angled, sometimes up to 1 m high; Leaves petioled, ovate, toothed or sinuately repand or entire; Flowers white, in drooping, sub-umbellate, extra-axillary cymes, white; Berries globose, dull black.

Flowering and Fruiting: July-October

348. Scientific Name: *Solanum torvum* Sw.

Synonym: *Solanum heterophyllum* Balb. *ex* Dunal, *nom. nud.*

Family: Solanaceae

Common Name: Turkey Berry

Vernacular Name: Bhurat

Habitat: Open tropical areas

Uses: Edible, Medicinal

IUCN Threat listing: Not Assessed (NA)

Description: Perennial, erect or spreading prickly shrub, 1 to 3 m tall; Stems initially green, brown on maturity; Leaves alternate, broadly ovate, 5-20 cm long, with broad, blunt lobes, stellately hairy, spined on main veins; Flowers in clusters, sepals finely hairy, five, petals white to cream, star-shaped, stamens 5, elongate, yellow; Fruits globular berries, at first green, ripening to dull yellow.

Flowering and Fruiting: July-March

349. Scientific Name: *Solanum viarum* Dunal

Synonym: *Solanum chloranthum* DC.

Family: Solanaceae

Common Name: Tropical Soda Apple

Vernacular Name: Kataiyyan

Habitat: Open tropical areas

Uses: Edible, Medicinal

IUCN Threat listing: Not Assessed (NA)

Description: Erect, perennial, 50-150 cm high; Stem short, pubescent, branches with recurved prickles up to 5 mm long; Leaves broadly ovate, up to 20 cm long, bluntly lobed, with markedly undulate edges, dark green, glossy above; Flowers white, in clusters of 1-5, 1.5 cm across; Fruits globose berry, mottled green when young, maturing yellow.

Flowering and Fruiting: August-April

350. Scientific Name: *Solanum virginianum* L.

Synonym: *Solanum surattense* Burm. *f.*

Family: Solanaceae

Common Name: Thorny Nightshade

Vernacular Name: Bhat Kataiyyan

Habitat: Open tropical areas

Uses: Edible, Medicinal

IUCN Threat listing: Not Assessed (NA)

Description: Prickly, deep-rooted, diffuse, biennial or perennial, herb/ undershrub; Stem woody at base, prickles yellow; Leaves ovate-oblong, stellately-hairy, pinnatifid, segments dentate, base rounded, armed with prickles; Flowers in 1-few-flowered cymes, bluish-purple, pedicels curved; Berries globose, nearly 2 cm in diameter, yellow.

Flowering and Fruiting: August-November

351. Scientific Name: *Sonchus oleraceus* L.

Synonym: *Sonchus gracilis* Phil.

Family: Asteraceae

Common Name: Common Sowthistle

Vernacular Name: Dudhi

Habitat: Open tropical areas

Uses: Medicinal

IUCN Threat listing: Not Assessed (NA)

Description: Annual or biennial herb, 40-150 cm tall; Leaves large, alternate, soft, glabrous and dull-green, with small, soft teeth,having two pointed auricles at the limb base, large; Flowers heads 20-50, yellow, each with 80-250 flowers which are longer than the involucres; Achenes brown, oblanceolate, transversely tuberculate-rugose.

Flowering and Fruiting: July-September

352. Scientific Name: *Spathodea campanulata* Beauv.

Synonym: *Bignonia tulipifera* Schum.

Family: Bignoniaceae

Common Name: African Tuliptree

Vernacular Name: Rugtoora

Habitat: Open tropical areas

Uses: Avenue tree

IUCN Threat listing: Not Assessed (NA)

Description: Large upright tree, with stout, tapering, somewhat buttressed trunk, having warty light gray bark; Leaves opposite, glossy deep green, pinnate, 1-2 ft long, clustered at end of branches; Flowers in clusters, cup-shaped when open, large, showy reddish orange in colour; Fruits green brown, fingerlike pods pointing upwards and outwards above the foliage.

Flowering and Fruiting: Throughout the year

353. Scientific Name: *Spermacoce articularis* L. *f.*

Synonym: *Borreria articularis* (L. *f.*) F.N.Williams

Family: Rubiaceae

Common Name: Jointed Buttonweed

Habitat: Open tropical areas

Uses: Medicinal

IUCN Threat listing: Not Assessed (NA)

Description: Procumbent, hispidly-hairy herb; Stem profusely branched, branches quadrangular; Leaves nearly sessile, decussate, ovate-oblong or elliptic-oblong, rigidly coriaceous, leathery; Flowers 4-6 at nodes, pale blue or pale violet in colour, tertramerous, shortly-pedicelled; Capsules oblong, hairy, with crown of persistent sepals.

Flowering and Fruiting: August-January

354. Scientific Name: *Spermacoce hispida* L.

Synonym: *Borreria hispida* (L.) K. Schum., *nom. illeg.*

Family: Rubiaceae

Common Name: Shaggy Buttonweed

Vernacular Name: Madanghanti

Habitat: Open tropical areas

Uses: Medicinal

IUCN Threat listing: Not Assessed (NA)

Description: A procumbent, branched, hairy or rough herb, 10 to 14 cm long; Leaves ovate, spathulate or elliptic, pointed or rounded at the tip; Flowers in axillary whorls of 4-6, calyx-teeth linear-lanceolate, corolla pale blue or white; Fruit a hairy capsule, about 5 mm long.

Flowering and Fruiting: May-August

355. Scientific Name: *Spermacoce ocymoides* Burm. *f.*

Synonym: *Borreria ocymoides* (Burm. *f.*) DC.

Family: Rubiaceae

Common Name: Purple Leaved Buttonweed

Habitat: Open tropical areas

Uses: Medicinal

IUCN Threat listing: Not Assessed (NA)

Description: A prostrate or spreading annual herb; Leaves broadly elliptic, either hairless or with scattered hairs on veins below and margin; Flower heads appear at the end of branches or in leaf axils, flowers white, petals much longer than the tube.

Flowering and Fruiting: October-December

356. Scientific Name: *Sphaeranthus indicus* L.

Synonym: *Sphaeranthus hirtus* Willd.

Family: Asteraceae

Common Name: East Indian Globe Thistle

Vernacular Name: Gorakhmundi

Habitat: Open tropical areas

Uses: Medicinal

IUCN Threat listing: Least Concerned (LC)

Description: Softly-hairy, sub-erect, strongly scented annual herb; Stems long, spreading, winged with decurrent leaf bases; Leaves obovate-oblong, narrowed to the base, dentate or serrate, glandular-hairy; Heads compound, terminal, leaf-opposed, globose, flowers numerous, small, crowded, pink or light purple; Achenes stalked, pappus absent.

Flowering and Fruiting: November-January

357. Scientific Name: *Spilanthes acmella* (L.) Murray

Synonym: *Bidens acmella* (L.) Lam.

Family: Asteraceae

Common Name: Toothache Plant

Vernacular Name: Akarkara

Habitat: Humid tropical areas

Uses: Edible, Medicinal

IUCN Threat listing: Not Assessed (NA)

Description: An evergreen herb of moist places, with prostrate or erect, often reddish, hairless stem; Leaves broadly ovate to triangular, 5-11 cm long, margins toothed, tip sharp; Flower-heads arise singly, elongated-conical, containing primarily disc florets, florets yellow to orange; Achenes are black, 2-2.5 mm long.

Flowering and Fruiting: April-November

358. Scientific Name: *Spondias pinnata* (L. *f.*) Kurz

Synonym: *Mangifera pinnata* L. *f.*

Family: Anacardiaceae

Common Name: Wild Mango

Vernacular Name: Ameda

Habitat: Humid tropical areas

Uses: Edible, Medicinal

IUCN Threat listing: Not Assessed (NA)

Description: A deciduous tree, up to 10-15 m tall; Leaves compound, 30-40 cm long, leaflets 5-11, opposite, stalked, ovate oblong-elliptic; Flowers tiny, borne in panicles at the end of branches, stalkless, white; Fruit is ellipsoid to elliptic-ovoid, yellowish orange at maturity.

Flowering and Fruiting: May-December

359. Scientific Name: *Stachytarpheta jamaicensis* (L.) Vahl

Synonym: *Verbena jamaicensis* L.

Family: Verbenaceae

Common Name: Blue Porterweed

Vernacular Name: Kariyatharini

Habitat: Humid tropical areas

Uses: Edible, Medicinal

IUCN Threat listing: Not Assessed (NA)

Description: A perennial herb, up to 120 cm; Stems sometimes woody at the base, hairy at the nodes, sometimes purplish; Leaves elliptic, oblong, obovate or spathulate, often decurrent into the petiole, hairless or with sparse hairs on the veins beneath; margin crenate-dentate; Calyx tubular, ellipsoid, 4-toothed with 2 central teeth shorter, corolla blue or violet, rarely almost white, tube 8-11 mm long.

Flowering and Fruiting: Throughout the year

360. Scientific Name: *Stephania japonica* (Thunb.) Miers

Synonym: *Menispermum japonicum* Thunb.

Family: Menispermaceae

Common Name: Tape Vine

Habitat: Humid tropical areas

Uses: Medicinal, Insecticidal

IUCN Threat listing: Not Assessed (NA)

Description: A slender creeper, up to 10 m in length; Leaves opposite, long petioled, papery, margin entire, primary veins 8-10; Inflorescence compound umbelliform cyme, axillary; Male flowers pentamerous, sepals yellowish green, petals slightly fleshy, yellow, female flowers trimerous; Fruits sub-globose, slightly compressed, nearly sessile drupes, shining, glabrous, red when ripe.

Flowering and Fruiting: April-December

361. Scientific Name: *Stereospermum colais* (Buch.-Ham. *ex* Dillw.) D. L. Mabberle

Synonym: *Stereospermum tetragonum* DC.

Family: Bignoniaceae

Common Name: Yellow Snake Tree

Vernacular Name: Paroli

Habitat: Open tropical areas

Uses: Avenue tree, Medicinal

IUCN Threat listing: Not Assessed (NA)

Description: A large deciduous tree, often buttressed up to 20 m in height; Leaves 30-45 cm long, imparipinnate, opposite, leaflets 5-11 in number, purple when young; Flowers bisexual, borne in drooping, large, glabrous panicles, yellow or dark purple with red coloured veins; Fruit spirally twisted, 4-angled capsule.

Flowering and Fruiting: January-August

362. Scientific Name: *Synedrella nodiflora* (L.) Gaertn.

Synonym: *Verbesina nodiflora* L.

Family: Asteraceae

Common Name: Cinderella Weed

Habitat: Humid tropical areas

Uses: Medicinal

IUCN Threat listing: Not Assessed (NA)

Description: Erect, branched, ephemeral herb, usually 30-80 cm tall; Leaves in opposite pairs, elliptic to ovate with 3 prominent veins, petioles short, margins finely toothed margins; Flowers in small crowded bunches of 2-8 heads at nodes, heads with several erect bracts, ray florets 5-6, disc florets 10-20, petals yellow.

Flowering and Fruiting: July-January

363. Scientific Name: *Syzygium cumini* (L.) Skeels

Synonym: *Eugenia cumini* (L.) Druce

Family: Myrtaceae

Common Name: Black Plum

Vernacular Name: Jamun

Habitat: Open tropical areas

Uses: Edible, Medicinal, Antimicrobial

IUCN Threat listing: Not Assessed (NA)

Description: Medium to large, evergreen tree; Stem with light-grey, smooth bark and terete branches; Leaves variable in shape, oblong-ovate, shortly acuminate, shining, glabrous, gland-dotted; Flowers tetramerous, small, greenish, in trichotomous lateral panicals; Fruit a berry, oblong-obovoid or sub-globose, pink while ripening, black purple and juicy when fully ripe.

Flowering and Fruiting: July-September

364. Scientific Name: *Tabernaemontana divaricata* (L.) R. Br. *ex* Roem. & Schult.

Synonym: *Ervatamia divaricata* (L.) Burkill

Family: Apocyanaceae

Common Name: Crape Jasmine

Vernacular Name: Tagar

Habitat: Humid tropical areas

Uses: Ornamental, Medicinal, Dye

IUCN Threat listing: Not Assessed (NA)

Description: Evergreen shrub, up to 5-6 ft. high; Leaves large, shiny, deep green; Flowers borne in small clusters at apices of stem, shining waxy, white, petals 5.

Flowering and Fruiting: July-September

365. Scientific Name: *Tagetes erecta* L.

Synonym: *Tagetes major* Gaertn.

Family: Asteraceae

Common Name: Marigold

Vernacular Name: Genda

Habitat: Open tropical areas

Uses: Ornamental, Aromatic, Medicinal

IUCN Threat listing: Not Assessed (NA)

Description: An erect, annual herb, up to 180 cm high; stems angular, glabrous; Leaves pinnate, aromatic on crushing; Flower heads 5 cm across and composed of a dense arrangement of rays, yellow, orange and a unique bronze color.

Flowering and Fruiting: September-November

366. Scientific Name: *Tagetes patula* L.

Synonym: *Tagetes corymbosa* Sweet

Family: Asteraceae

Common Name: French Marigold

Vernacular Name: Genda

Habitat: Open tropical areas

Uses: Ornamental, Aromatic, Medicinal

IUCN Threat listing: Not Assessed (NA)

Description: An annual, occasionally reaching 50 cm in height; Leaves simple, alternate and oppositely arranged, the edge of leaf blade has lobes; Flower heads orange-red coloured, strap-shaped, have tubular disk flowers in center of ray flowers.

Flowering and Fruiting: September-November

367. Scientific Name: *Tamarindus indica* L.

Synonym: *Tamarindus officinalis* Hook.

Family: Leguminosae

Common Name: Tamarind

Vernacular Name: Imli

Habitat: Open tropical areas

Uses: Edible, Medicinal, Furniture

IUCN Threat listing: Not Assessed (NA)

Description: Large, deciduous tree; Trunk massive, bark dark-grey, fissured, branches drooping branches, umbrella-shaped crown; Leaves pinnately-compound, leaflets opposite, 10-18 pairs, oblong; Flowers borne in small drooping clusters, pale yellow with purple or red veins; Pods velvety, deep brown, sausage shaped, constricted between the seeds.

Flowering and Fruiting: May-October

368. Scientific Name: *Tecoma stans* (L.) Juss. *ex* Kunth

Synonym: *Bignonia stans* L.

Family: Bignoniaceae

Common Name: Yellow Bells

Vernacular Name: Piliya

Habitat: Open tropical areas

Uses: Avenue tree, Medicinal, Timber

IUCN Threat listing: Not Assessed (NA)

Description: Perennial shrub, usually 1.5-5 m high; Stems smooth, hairless and green when young, pale brown or reddish-brown with age; Leaves 10-25 cm long, in pairs, stalks slender, 1-9 cm long; Flowers showy, bright yellow, borne in several-flowered clusters, tubular in shape.

Flowering and Fruiting: Throughout the year

369. Scientific Name: *Tectona grandis* L. *f.*

Synonym: *Jatus grandis* (L. *f.*) Kuntze

Family: Lamiaceae

Common Name: Teak

Vernacular Name: Sagon

Habitat: Open tropical areas

Uses: Timber, Furniture, Dye, Medicinal

IUCN Threat listing: Not Assessed (NA)

Description: Large, deciduous tree; Main trunk massive, with thin, light-brown or grey bark, branches 4-angled; Leaves simple, opposite, leathery, entire, ovate, elliptic or obovate, apex acute, base rounded; Flowers white, in di-trichotomously branched cyme, sweet-scented; Drupes sub-globose, covered with hairs, enveloped loosely inside the flattened, enlarged calyx.

Flowering and Fruiting: May-January

370. Scientific Name: *Terminalia alata* Heyne *ex* Roth

Synonym: *Terminalia macrocarpa* Steud.

Family: Combretaceae

Common Name: Indian Laurel

Vernacular Name: Saja

Habitat: Open tropical areas

Uses: Timber, Furniture

IUCN Threat listing: Not Assessed (NA)

Description: Large deciduous tree; Stem with dark grey or blackish, deeply cracked bark, young parts rusty-tomentose; Leaves nearly opposite, smooth above, woolly beneath, base often unequal; Flowers pale or greenish-yellow, in spikes, petals absent; Fruits 5-winged, 3-5 cm long.

Flowering and Fruiting: April-June

371. Scientific Name: *Terminalia arjuna* (Roxb.) Wight & Arn.

Synonym: *Terminalia ovalifolia* Rottl. *ex* C.B. Cl.

Family: Combretaceae

Common Name: Arjun Tree

Vernacular Name: Kahuva

Habitat: Open tropical areas

Uses: Medicinal

IUCN Threat listing: Not Assessed (NA)

Description: Large tree; Stem with smooth grey bark, peeling off in large flakes; Leaves nearly opposite, oblong or elliptic, unequal at the base, slightly acute at the apex; Flowers small, sessile, greenish-yellow, in short, axillary spikes, or terminal, pendulous panicles; Fruits ovoid or obovoid-oblong, woody, with 5-7, hard, coriaceous wings.

Flowering and Fruiting: April-July

372. Scientific Name: *Terminalia bellirica* (Gaertn.) Roxb.

Synonym: *Terminalia belirica* (Gaertn.) Wall.

Family: Combretaceae

Common Name: Beach Almond

Vernacular Name: Baheda

Habitat: Open tropical areas

Uses: Avenue tree, Fodder, Medicinal

IUCN Threat listing: Not Assessed (NA)

Description: Large, deciduous tree; Stem with bluish-grey bark; Leaves alternate, crowded towards ends of branches, elliptic or ovbovate, acuminate, unequal at base; Flowers in interrupted, slender spikes borne in the axils of leaves, sessile, greenish-yellow; Fruit an ovoid-ellipsoid or globose drupe, grey-velvety.

Flowering and Fruiting: December-March

373. Scientific Name: *Terminalia chebula* Retz.

Synonym: *Myrobalanus chebula* (Retz.) Gaertn.

Family: Combretaceae

Common Name: Chebulic Myrobalan

Vernacular Name: Harra

Habitat: Open tropical areas

Uses: Medicinal

IUCN Threat listing: Not Assessed (NA)

Description: Large deciduous tree; Stem with dark-brown bark, young parts rusty-tomentose; Leaves mostly sub-opposite, 8-15 cm long, ovate-oblong, acuminate, base acute or rounded; Flowers bisexual, sessile, dull-white, scented, in terminal, often panicled spikes; Fruit 3-5 cm long, 5-ribbed, ovoid.

Flowering and Fruiting: February-August

374. Scientific Name: *Tetrastigma leucostaphylum* (Dennst.) A. Alston

Synonym: *Tetrastigma lanceolarium* (Roxb.) Planch.

Family: Vitaceae

Common Name: Indian Chestnut Vine

Vernacular Name: Ambrolla

Habitat: Humid tropical areas

Uses: Edible, Medicinal

IUCN Threat listing: Not Assessed (NA)

Description: Woody scandent creeper; stems strongly verrucose, with elongated lenticels, tendrils simple, woody with age; Leaves simple or pedately 3-5-foliolate, sometimes all on same plant, leaflets oblong-lanceolate, base cuneate, margin coarsely serrate, especially distally, apex acute, coriaceous; Flowers in short axillary cymes, calyx lobes 4, ciliolate, petals 4, white, ovate, acute, apically thickened; Berry sub-globose, creamy white, seeds 2-4, oblong.

Flowering and Fruiting: May-August

375. Scientific Name: *Themeda quadrivalvis* (L.) Kuntze

Synonym: *Andropogon quadrivalvis* L.

Family: Poaceae

Common Name: Kangaroo Grass

Habitat: Open tropical grasslands

Uses: Fodder

IUCN Threat listing: Not Assessed (NA)

Description: An upright, short-lived grass, 50-200 cm tall; Leaves linear, base sub-cordate, ligule forms scarious rim, sheaths glabrous, keeled, purple-tinged; Involucral spikelets lanceolate, sharply acute, lower glumes lanceolate, acuminate.

Flowering and Fruiting: September-December

376. Scientific Name: *Thespesia lampas* (Cav.) Dalzell & A. Gibson

Synonym: *Azanza lampas* (Cav.) Alef.

Family: Malvaceae

Common Name: Common Mallow

Vernacular Name: Ban Kapas

Habitat: Humid tropical areas

Uses: Medicinal, Fibre

IUCN Threat listing: Not Assessed (NA)

Description: Erect, slightly branched shrub, 2-3 m high; Leaves ovate, 10-20 cm long, somewhat 3-lobed or nearly entire, smooth above, somewhat hairy below, broad-cordate at base, tip pointed; Flowers large, borne in leaf axils of terminal branches, usually in threes, calyx green, 5 sepals united below the middle, corolla bell-shaped, yellow, dark-purple at the center; Capsules ovoid, about 3 cm long, 4-5-valved.

Flowering and Fruiting: August-December

377. Scientific Name: *Thespesia populnea* (L.) Soland. *ex* Correa

Synonym: *Hibiscus populneus* L.

Family: Malvaceae

Common Name: Indian Tulip Tree

Vernacular Name: Paras Pipal

Habitat: Humid tropical areas

Uses: Furniture, Fibre

IUCN Threat listing: Not Assessed (NA)

Description: Medium-sized tree, usually 6-9 m tall, crown dense, evergreen; Leaves alternate, spirally arranged, blade orbicular, deltoid, ovate or oblong, simple and entire; Flowers solitary in leaf axils, bisexual, opening one at a time, on stout pedicels, calyx cup-shaped, green, petals 5, broad rounded oblique, pale yellow, usually with maroon spot at base; Fruit rounded but flattened, about 3 cm in diameter.

Flowering and Fruiting: March-June

378. Scientific Name: *Thunbergia erecta* (Benth.) T. Anders.

Synonym: *Meyenia erecta* Benth.

Family: Acanthaceae

Common Name: King's Mantle

Habitat: Open tropical areas

Uses: Ornamental

IUCN Threat listing: Not Assessed (NA)

Description: An upright shrub, up to 4 feet; Leaves opposite, ovate-elliptic, margin entire or wavy, occasionally with a broad triangular tooth above the middle; Flowers in leaf axils, singly or in pairs, sepal cup is short, bowl shaped, flower tube flares open into five dark purple, roundish petals, yellowish-white at the base.

Flowering and Fruiting: August-January

379. Scientific Name: *Thunbergia fragrans* Roxb.

Synonym: *Thunbergia laevis* Nees.

Family: Acanthaceae

Common Name: Whitelady

Vernacular Name: Chimin

Habitat: Humid tropical areas

Uses: Ornamental, Medicinal

IUCN Threat listing: Not Assessed (NA)

Description: A climbing herb, up to 6 m tall, leaves are scabrid; oppositely arranged; entire and obovate with cuneate tip, Stems are usually green or reddish coloured; quadrangular in cross section and hairy; Tubular white flowers are borne singly or in pairs; Fruit is capsular 1 cm across; it is having long thin beak upto 2 cm.

Flowering and Fruiting: January-May

380. Scientific Name: *Thysanolaena latifolia* (Roxb. *ex* Hornem.) Honda

Synonym: *Thysanolaena maxima* (Roxb.) Kuntze

Family: Poaceae

Common Name: Broom Grass

Vernacular Name: Jhadu Ghans

Habitat: Humid tropical areas

Uses: Fodder, Hedge plant, Erosion control

IUCN Threat listing: Not Assessed (NA)

Description: Evergreen perennial grass, growing to 3 m; Leaves large, up to 7 cm wide; Inflorescence an oblong-ovate panicle, 30-60 cm, spikelets minute,lanceolate, laterally compressed, comprising 1 basal sterile floret without conspicuous palea, 1 fertile floret and a barren rachilla extension, lodicules 2, ovary glabrous; Fruit caryopsis with adherent pericarp.

Flowering and Fruiting: March-June

381. Scientific Name: *Tithonia diversifolia* (Hemsl.) A. Gray

Synonym: *Mirasolia diversifolia* Hemsl.

Family: Asteraceae

Common Name: Mexican Sunflower

Habitat: Open tropical areas

Uses: Ornamental

IUCN Threat listing: Not Assessed (NA)

Description: Robust, perennial herbaceous plant, up to 3 m; Leaves opposite or alternate, sub-ovate, rough, hairy, 3-7-lobed, base attenuate or decurrent, apex acute or acuminate, margin crenate; Flower large, showy, yellow, in solitary heads, peduncle 6-13 cm long.

Flowering and Fruiting: June-December

382. Scientific Name: *Toona ciliata* M. Roem.

Synonym: *Cedrela toona* Roxb. *ex* Rottl. & Willd.

Family: Meliaceae

Common Name: Toon Tree

Vernacular Name: Toon

Habitat: Open tropical areas

Uses: Medicinal, Dye

IUCN Threat listing: Least Concerned (LC)

Description: Large, deciduous tree; Crown dense, spreading, bark dark grey, branching high above the ground; Leaves paripinnate, glabrous, leaflets 8-30, usually opposite, entire or undulate, serrate, oblique at the base; Flowers white/cream, mildly fragrant, in short, terminal, drooping panicles; Capsules oblong or oblanceolate, dark brown in colour, smooth, dehiscing by 5-valves.

Flowering and Fruiting: February-May

383. Scientific Name: *Tradescantia pallida* (Rose) D.R. Hunt

Synonym: *Setcreasea pallida* Rose

Family: Commelinaceae

Common Name: Purple Heart

Habitat: Open tropical areas

Uses: Ornamental

IUCN Threat listing: Not Assessed (NA)

Description: An evergreen, perennial, scrambling plant; Stems fleshy, erect at first, later lying on the ground as a creeping herb; Leaves lance-shaped, covered with pale hairs; Flowers emerge from curving double bracts at the stem tips, pale orchid-pink, 3-petaled.

Flowering and Fruiting: Throughout the year

384. Scientific Name: *Trema orientalis* (L.) Bl.

Synonym: *Celtis orientalis* L.

Family: Cannabaceae

Common Name: Indian Charcoal Tree

Vernacular Name: Jivan

Habitat: Open tropical areas

Uses: Fuel wood, Fibre, Paper making, Medicinal

IUCN Threat listing: Not Assessed (NA)

Description: Small, evergreen tree; Stem with dark grey, thin bark, branches spreading, straight; ovate-oblong, membraneous, serrate, scarbid above, soft tomentose beneath, acuminate, base cordate; Flowers in lax axillary cymes, yellow; Drupes ovoid, black when ripe.

Flowering and Fruiting: September-December

385. Scientific Name: *Trichodesma zeylanicum* (Burm. *f.*) R. Br.

Synonym: *Boraginella zeylanica* (Burm. *f.*) Kuntze

Family: Boraginaceae

Common Name: Camel Bush

Habitat: Open tropical areas

Uses: Medicinal

IUCN Threat listing: Not Assessed (NA)

Description: An erect herb, up to 1 m high, stems and twigs clothed in white scabrous hairs; Leaf blades very variable in size, with white scabrous hairs on both sides; Flowers blue to occasionally white, sepals clothed in white appressed hairs on both the inner and outer surfaces, corolla cream towards the centre but the lobes blue, ending in long spirally coiled tips; Nutlets about 3-4 mm long.

Flowering and Fruiting: August-November

386. Scientific Name: *Trichosanthes tricuspidata* Lour.

Synonym: *Trichosanthes palmata* Roxb.

Family: Cucurbitaceae

Common Name: Indrayan

Habitat: Humid tropical areas

Uses: Medicinal, Antimicrobial

IUCN Threat listing: Not Assessed (NA)

Description: Large climber, 5-20 m long; Stem strong, woody and grooved, with trailing branches; Leaves, broadly heart-shaped, but palmately 3-5 lobed, leaf margin serrated; Flowers unisexual, occur either singly or in 5-10 flowered racemes, in leaf axils, white, petals 5, wedge-shaped with frilly margins; Fruit spherical, 4-5 cm across, red when ripe.

Flowering and Fruiting: April-August

387. Scientific Name: *Tridax procumbens* L.

Synonym: *Chrysanthemum procumbens* (L.) Sessé & Moc.

Family: Asteraceae

Common Name: Tridax Daisy

Vernacular Name: Khal Muriya

Habitat: Open tropical areas

Uses: Antimicrobial, Insecticidal, Wound healing, Medicinal

IUCN Threat listing: Not Assessed (NA)

Description: Weak annual; Stem straggling or prostrate, 30-60 cm high; Leaves few, petioled, ovate-lanceolate, toothed, hispid above and beneath, base acute; Flowers creamy-yellow, heads on very long, solitary, pilose peduncles, ray florets female, disk florets bisexual; Achenes with silky hairs, dark brown, pappus feathery.

Flowering and Fruiting: Throughout the year

388. Scientific Name: *Triumfetta rhomboidea* Jacq.

Synonym: *Bartramia indica* L.

Family: Malvaceae

Common Name: Burr Bush

Vernacular Name: Chikti

Habitat: Open tropical areas

Uses: Medicinal, Fibre

IUCN Threat listing: Not Assessed (NA)

Description: An erect, pubescent, hairy herb/undershrub; Stem much branched from the base; Leaves variable in shape, broadly ovate, pubescent on both the surfaces, base variable; Flowers yellow, in dense, terminal and leaf-opposed cymes; Fruits globular, clothed with smooth, hooked spines.

Flowering and Fruiting: August-January

389. Scientific Name: *Turnera ulmifolia* L.

Synonym: *Turnera cuneifolia* Juss. *ex* Urb.

Family: Turneraceae

Common Name: Yellow Alder

Habitat: Open tropical areas

Uses: Ornamental, Medicinal

IUCN Threat listing: Not Assessed (NA)

Description: Small, annual herb; Leaves clustered toward the tips of the branches, lanceolate to oblong-lanceolate; Flowers in terminal clusters, calyx tube 5-8 mm long, the lobes lanceolate, apex caudate, petals yellow, sometimes with a brown spot toward base, obovate, 20-35 mm long; Capsules ovoid, 7-8 mm long.

Flowering and Fruiting: September-November

390. Scientific Name: *Typha angustifolia* L.

Synonym: *Typha minor* Curtis

Family: Typhaceae

Common Name: Narrow Leaf Cattail

Vernacular Name: Patera

Habitat: Humid tropical areas

Uses: Phytoremediation

IUCN Threat listing: Least Concerned (LC)

Description: Perennial, robust, unbranched grass, 1-2 m high; Leaves linear, green, glabrous and rather stiff, the inner side flat to slightly concave, outer side convex; Flowers in spikes, staminate spike above the pistillate spike, separated by at least half inch, staminate spike light yellow-light brown, the mature pistillate spike brown or reddish brown; Fertile pistillate florets mature into achenes.

Flowering and Fruiting: December-February

391. Scientific Name: *Uraria lagopodoides* (L.) DC

Synonym: *Uraria lagopoides* (Burm. *f.*) DC.

Family: Leguminosae

Common Name: Hare Foot Uraria

Vernacular Name: Pithran

Habitat: Open tropical areas

Uses: Medicinal, Insecticidal

IUCN Threat listing: Not Assessed (NA)

Description: A prostrate or spreading herb, up to 2 ft tall; Leaves mostly trifoliolate, rarely 1-foliolate, stalk 1-2 cm long, central leaflet nearly round or elliptic-ovate, gray-yellow velvety beneath; Flowers pale purple, borne in densely flowered racemes, 3-6 cm long, at the end of branches, standard obovate, base flat; Pod is enclosed by sepal cup, black-brown at maturity, small.

Flowering and Fruiting: May -November

392. Scientific Name: *Uraria picta* (Jacq.) DC.

Synonym: *Hedysarum pictum* Jacq.

Family: Leguminosae

Common Name: Uraria

Vernacular Name: Dabran

Habitat: Open tropical areas

Uses: Medicinal

IUCN Threat listing: Least Concerned (LC)

Description: An erect, little branched, perennial herb upto 1.8 m long; Leaves 20-30 cm, imparipinnate, leaflets 5-9, linear oblong, acute, blotched with white, minutely pubescent below, often variegated, base rounded, shiny and hairless above; Inflorescence a long terminal densely many-flowered spike, flowers pink, blue or reddish; Fruit 5-9 mm long, folded into 3 segments, brown to black in colour.

Flowering and Fruiting: April-July

393. Scientific Name: *Urena lobata* L.

Synonym: *Malachra urena* DC.

Family: Malvaceae

Common Name: Caesarweed

Vernacular Name: Lapetua

Habitat: Open tropical areas

Uses: Edible, Medicinal, Fibre

IUCN Threat listing: Not Assessed (NA)

Description: Undershrub, 0.6-1.2 m high; Stem tomentose, with stellate hairs; Leaves variable in shape, round or ovate or 5-7 lobed, broader than long, base cordate, tomentose; Flowers in axillary clusters, with very small pedicels, pink with dark centre; Fruit a capsule, stellately hairy, hooked.

Flowering and Fruiting: Throughout the year

394. Scientific Name: *Urochloa panicoides* P. Beauv.

Synonym: *Panicum panicoides* (P. Beauv.) Hitchc.

Family: Poaceae

Common Name: Liverseed Grass

Vernacular Name: Kuriya

Habitat: Open tropical grasslands

Uses: Fodder

IUCN Threat listing: Least Concerned

Description: Annual grass, stems decumbent or upright, up to a 1 m long; Leaves up to 25 cm long, blade linear or lance-shaped, hairless to somewhat hairy; Inflorescence of up to 10 racemes, each up to 7 cm long, the spikelets solitary or paired, glumes unequal, the lower few as long as the spikelet, upper lemma mucronate.

Flowering and Fruiting: August-November

395. Scientific Name: *Utricularia aurea* Lour.

Synonym: *Utricularia pilosa* (Makino) Makino

Family: Lentibulariaceae

Common Name: Golden Bladderwort

Habitat: Humid tropical areas

Uses: Ornamental, Wound healing

IUCN Threat listing: Least Concerned (LC)

Description: Submerged, perennial or annual, aquatic herb; Leaves alternately arranged, very finely pinnately compound, utricles located at the leaf segments, obliquely ovoid in shape; Inflorescence erect, 5-25 cm long, flowers yellow, ovary globose, glandular, style long, stigma bi-lipped, ciliate; Capsule globose, up to 5 mm long, relatively thick and fleshy, circumscissile, seeds numerous.

Flowering and Fruiting: May-July

396. Scientific Name: *Vachellia farnesiana* (L.) Wight & Arn

Synonym: *Acacia farnesiana* (L.) Willd.

Family: Leguminosae

Common Name: Mimosa Bush

Vernacular Name: Gandh Babool

Habitat: Open tropical areas

Uses: Avenue tree, Aromatic, Medicinal

IUCN Threat listing: Not Assessed (NA)

Description: Shrub/small tree; Branches slender, zigzag, with grey dots; Leaves bipinnate, pinnae 4-8 pairs, leaflets 10-20 pairs, linear-oblong, stipular spines straight and white; Flowers in heads, bright yellow; Pods 5-8 cm long, thick, cylindrical, horizontally striate.

Flowering and Fruiting: May-November

397. Scientific Name: *Vachellia nilotica* (L.) P.J.H. Hurter & Mabb.

Synonym: *Acacia nilotica* (L.) Willd. *ex* Delile

Family: Leguminosae

Common Name: Gum Arabic

Vernacular Name: Babool

Habitat: Open tropical areas

Uses: Fodder, Gum, Timber

IUCN Threat listing: Not Assessed (NA)

Description: Moderate/large tree; Crown spreading, bark dark black, having large longitudinal cracks; Pinnae 3-6 pairs, leaflets 10-20 pairs, membraneous, glabrous, spines straight; Flowers yellow, in heads; Pods usually solitary, 7-15 cm long, stalked, fleshy, sutures deeply indented between the seeds.

Flowering and Fruiting: July-December

398. Scientific Name: *Verbascum coromandelianum* (L.) D. McKean

Synonym: *Verbascum chinense* (L.) Li

Family: Scrophulariaceae

Common Name: Chinese Mullein

Vernacular Name: Kokhima

Habitat: Open tropical areas

Uses: Medicinal

IUCN Threat listing: Not Assessed (NA)

Description: Erect, pubescent annual herb; Stem branched from the base/unbranched, up to 1.5 m high, glandular-pubescent on upper portion; Radical leaves petioled, crowded, lyrate-pinnatifid, stem leaves small, sessile; Flowers yellow, sessile, in simple or branched terminal racemes; Capsules subglobose, glandular.

Flowering and Fruiting: June-August

399. Scientific Name: *Vernonia cinerea* (L.) Less.

Synonym: *Cyanthillium cinereum* (L.) H. Roxb.

Family: Asteraceae

Common Name: Little Ironweed

Vernacular Name: Sehdevi

Habitat: Open tropical areas

Uses: Medicinal

IUCN Threat listing: Not Assessed (NA)

Description: An erect, variable annual; Stem terete, pubescent, slightly branched; Leaves nearly sessile, lanceolate or ovate, 1.5-3.5 cm, pubescent on both sides, teeth few; Heads numerous, homogamous, pink to purple-violet, in corymbs; Achenes slightly obovate, flattened, base conical, pappus white.

Flowering and Fruiting: Throughout the year

400. Scientific Name: *Vinca rosea* (L.) G. Don

Synonym: *Vinca rosea* L.

Family: Apocyanaceae

Common Name: Periwinkle

Vernacular Name: Ṣadabahar

Habitat: Open tropical areas

Uses: Ornamental, Medicinal

IUCN Threat listing: Not Assessed (NA)

Description: An evergreen sub-shrub or herb, up to 1 m tall; Leaves opposite, oval to oblong, glossy green, hairless, with a pale midrib and short petiole; Flowers white to dark pink with darker red centre, basal tube 2.5–3 cm long, corolla 2–5 cm across; Fruit a pair of follicles, 2–4 cm long.

Flowering and Fruiting: All year round

401. Scientific Name: *Vitex negundo* L.

Synonym: *Agnus-castus negundo* (L.) Carrière

Family: Lamiaceae

Common Name: Chaste Tree

Vernacular Name: Nirgundi

Habitat: Open tropical areas

Uses: Medicinal, Insecticidal

IUCN Threat listing: Not Assessed (NA)

Description: Deciduous shrub; Stem up to 2.5 m high, branches 4-angled, tomentose; Leaves digitately 3-5-foliolate, leaflets stalked, lanceolate, serrate or entire; Flowers purplish, pedicelled, in terminal panicles, up to 30 cm long; Fruit a succulent drupe, globose, enclosed in accrescent calyx, black when ripe.

Flowering and Fruiting: June-December

402. Scientific Name: *Volkameria inermis* L.

Synonym: *Clerodendrum inerme* (L.) Gaertn.

Family: Lamiaceae

Common Name: Glory Bower

Vernacular Name: Sankuppi

Habitat: Open tropical areas

Uses: Medicinal, Hedge species

IUCN Threat listing: Not Assessed (NA)

Description: A much branched, straggling shrub, 1-2 m tall; Leaves simple, opposite, ovate to elliptical, 5-10 cm long, tip acute to acuminate, green, smooth, margins entire; Flowers in cymes or umbels, usually 3 flowers joined at base, petals 5, white, tube long, narrow, stamens 4, purple, protruding.

Flowering and Fruiting: August-January

403. Scientific Name: *Wendlandia heynei* (Schult.) Santapau & Merchant

Synonym: *Wendlandia exserta* (Roxb.) DC.

Family: Rubiaceae

Common Name: Heyne's Wendlandia

Vernacular Name: Tilki

Habitat: Open tropical areas

Uses: Timber, Medicinal

IUCN Threat listing: Not Assessed (NA)

Description: Small, deciduous tree; Trunk with rough, brown, longitudinally cracked bark; Leaves lanceolate, acute, pubescent above; Flowers white, sub-sessile, fragrant, very numerous, in tomentose pyramidal panicles; Capsules long, grey tomentose.

Flowering and Fruiting: April-June

404. Scientific Name: *Woodfordia fruticosa* (L.) Kurz

Synonym: *Lythrum fruticosum* L.

Family: Lythraceae

Common Name: Fire Flame Bush

Vernacular Name: Dhawai

Habitat: Open tropical areas

Uses: Edible, Medicinal, Dye

IUCN Threat listing: Least Concerned (LC)

Description: Large shrub; Stem with long, spreading branches and reddish bark; Leaves sessile, lanceoalte, tapering to a fine point, upper surface green, lower whitish; Flowers in shortly-stalked clusters around branches, bright red; Capsules small, enclosed in the calyx.

Flowering and Fruiting: January-April

405. Scientific Name: *Xanthium strumarium* L.

Synonym: *Xanthium occidentale* Bertol.

Family: Asteraceae

Common Name: Cocklebur

Vernacular Name: Ghagra

Habitat: Open tropical areas

Uses: Antimicrobial

IUCN Threat listing: Not Assessed (NA)

Description: Scabrous, unarmed herb; Stem spotted, with short, rigid and few branches; Leaves petioled, triangular-cordate, unequal at base, undulate, irregularly toothed; Heads in terminal and axillary racemes, many, crowded at the top; Fruiting involucres ovoid or oblong, covered with hooked prickles.

Flowering and Fruiting: July-September

406. Scientific Name: *Zingiber zerumbet* (L.) Roscoe *ex* Sm.

Synonym: *Amomum zerumbet* L.

Family: Zingiberaceae

Common Name: Bittter Ginger

Vernacular Name: Kulanjan

Habitat: Humid tropical areas

Uses: Aromatic, Medicinal, Edible

IUCN Threat listing: Not Assessed (NA)

Description: A rhizomatous tuberous herb; Leaves sessile or short-petiolate, sheaths glabrescent, green, blades broadly lanceolate or oblong-lanceolate, acuminate at apex, cuneate at base; Inflorescence ellipsoid, calyx membranous, white, petals lanceolate, pale yellow to white, labellum pale yellow to white, the central lobe emarginate, the lateral lobes free nearly to base; Capsule ellipsoid or obovoid, red.

Flowering and Fruiting: July-November

407. Scientific Name: *Ziziphus mauritiana* Lam.

Synonym: *Rhamnus jujuba* L.

Family: Rhamnaceae

Common Name: Common Jujube

Vernacular Name: Boir

Habitat: Open tropical areas

Uses: Edible, Medicinal

IUCN Threat listing: Not Assessed (NA)

Description: Erect, medium-sized tree; Crown spreading, bark dark-grey to nearly black, young parts rusty tomentose; Leaves round or ovate, closely serrulate, green shining above, woolly tomentose below, prickles solitary or paired; Flowers greenish-yellow, in short axillary cymes or fascicles; Drupe variable in size and shape, red or orange-coloured when ripe.

Flowering and Fruiting: January-June

408. Scientific Name: *Ziziphus oenopolia* (L.) Mill.

Synonym: *Rhamnus oenopolia* L.

Family: Rhamnaceae

Common Name: Jackal Jujube

Vernacular Name: Jarwa Boir

Habitat: Open tropical areas

Uses: Edible, Medicinal

IUCN Threat listing: Not Assessed (NA)

Description: Thorny, straggling shrub; Branches sarmentose; Leaves obliquely ovate-lanceolate, obscurely crenate, rusty tomentose beneath, prickles recurved, solitary; Flowers in dichotomous, compresed cymes; Drupes subglobose, black when ripe.

Flowering and Fruiting: September-December

409. Scientific Name: *Ziziphus rugosa* Lam.

Synonym: *Rhamnus glabrata* Heyne

Family: Rhamnaceae

Common Name: Wild Jujube

Vernacular Name: Suran

Habitat: Open tropical areas

Uses: Edible, Medicinal

IUCN Threat listing: Not Assessed (NA)

Description: A small tree or straggling shrub, armed with stipular spines; Leaves broadly elliptic, short-tipped, denticulate, base oblique, wrinkled, petiole densely velvet-hairy; Inflorescence of pedunculate cymes, densely pubescent, calyx puberulous inside, with ovate acute lobes, petals absent, disc 5-lobed, styles 2, connate below the middle; Fruit globose or pyriform, white when ripe.

Flowering and Fruiting: November-May

410. Scientific Name: *Zornia gibbosa* Span.

Synonym: *Zornia cantonensis* Mohlenbr.

Family: Leguminosae

Common Name: Grass-like Zornia

Habitat: Open tropical areas

Uses: Medicinal

IUCN Threat listing: Not Assessed (NA)

Description: Grass-like, annual diffuse herb; Stem prostrate; Leaflets 2, shortly-petioled, punctate beneath, lanceolate, apex acute, base oblique-rounded; Flowers whitish-yellow, in lax spikes; Pods 3-6-jointed, densely prickly.

Flowering and Fruiting: September-January

■■■■

DISCUSSION

The study reported 410 species of flowering plants from the Mekal hill region, belonging to 321 genera, with genus to species ratio of 0.78. Since most of the flowering plants (around 70%) were discovered from Achanakmar-Amarkantak region, it is presumed that these sites have more conducive environment for speciation. This can be partly attributed to topography which ranges from 300-900 m, providing much elevation gradient, and also to the fact that these hills are responsible for orographic rainfall during monsoon. The region is also very important from hydrological point of view, as rivers like Narmada, Son, Arpa and Johila emerge from this area. Most of these rivers are rain-fed, but maintain their perennial nature owing to lush green forest of the region, which is able to retain moisture during dry seasons.

The species reported here show dominance of herbs (54%) probably, due to the availability of more open areas. Herbs are followed by trees (24%) mainly found near undisturbed forests or ecotones. Geologically, bulk of the soils have arisen out of igneous rocks and at places the depth of soil profile is shallow, such places are generally suited for herbs and early colonizers which prefer plenty of sunlight and less of nutrients. While in the forest, the biome dominated by tree species is having less of sunlight and more nutrients due to litter decomposition. Most dominant shrub in this region is *Lantana camara*, which is causing threat to native vegetation. This species spreads rapidly due to its massive seed production.

Phenology of flowering plants is directly and strongly correlated with prevailing seasons of the locality. Out of the four different seasons in the locality, summer and autumn seem to be phenologically most important, with 30% and 28% flowering/fruiting species, respectively. Most of the angiosperms either flower or fruit in this season, implying that the high temperature and moisture during the monsoon is important for blossoming of these plants. It could be said that moisture is important for 60% of plant species while 40% of the species blossom during winters or spring. Hence these plants may be influenced by temperature or day length.

Since assessment of the plant species was based on the sighting, the criteria for assessment of threat was based on IUCN red listing. It is global in nature and it gives accurate status of species population and threats it is facing. More than

85% of the species were found to be not assessed category (NA) hence, it could be concluded well that there is scanty data on their population. Since most of the earlier floristic works were confined to listing or documentation only, hardly any data on population is available. Autecological works are common especially for high value plants but synecology seems to be rather neglected aspect.

Some of the NA plant species are - *Abelmoschus moschatus, Abroma angusta, Baccharoides anthelmintica, Bambusa vulgaris, Caesalpinia decapetala, Caesalpinia pulcherrima, Dalbergia sissoo, Datura metel, Eichhornia crassipes, Elaeocarpus ganitrus* etc. While those in the least concerned (LC) category are - *Acacia auriculiformis, Aeschynomene indica, Bauhinia purpurea, Caesulia axillaris, Caryota urens, Dendrocalamus membranaceus, Dichrostachys cinerea, Echinochloa colona, Eleusine indica* etc. In data deficient (DD) category the species are - *Eclipta prostrata, Elaeodendron glaucum, Euphorbia milii* and *Mangifera indica*. While, only specie reported to be in vulnerable list (VU) is *Jacaranda mimosifolia.*

The species population trends are categorized into increasing, decreasing, stable and unknown threat categories. The species with increasing population are - *Alternanthera sessilis*, *Eleusine indica* and *Opuntia stricta,* facing least threats and are widely available. While the only species with decreasing population is *Jacaranda mimosifolia*, a native of Brazil and introduced in India for its ornamental value. While those in stable category are - *Acacia auriculiformis*, *Aeschynomene indica*, *Uraria picta*, *Urochloa panicoides*, *Bauhinia racemosa*, *Cajanus scarabaeoides*, *Cascabela thevetia*, *Commelina diffusa*, *Cyperus difformis*, *Cyperus rotundus*, *Dendrocalamus membranaceus*, *Dichrostachys cinerea* etc. While those species whose population status unknown are - *Echinochloa colona*, *Elaeodendron glaucum*, *Euphorbia milii*, *Fimbristylis quinquangularis*, *Grangea maderaspatana*, *Hygrophila schulli*, *Ipomoea aquatica*, *Lindernia dubia*, *Mangifera indica*, *Magnolia champaca*, *Ottelia alismoides* etc.

Some of the species facing threats from unknown sources are - *Acacia auriculiformis, Aeschynomene indica, Alternanthera sessilis, Bauhinia purpurea, Caesulia axillaris, Chionachne gigantea, Colocasia esculenta, Commelina diffusa, Cyperus difformis, Cyperus esculentus* etc. Hence, it could be said that either these species are abundantly available elsewhere or, a little information somehow very few is available on their population status. Species experiencing threat due to timber extraction are - *Magnolia champaca* and *Toona ciliata*. While species facing threats due to anthropogenic reasons are - *Merremia emarginata* and *Ottelia alismoides,* and the species under threat due to logging of natural forest are - *Caryota urens, Dendrocalamus hamiltonii, Dendrocalamus membranaceus* and *Epiphyllum oxypetalum*. Shifting cultivation, which is a primitive form of agriculture, also threatens both, the flora and fauna of the area. Some of the plants under threat due to the induced fire from this type of cultivation are - *Caryota urens, Delonix regia, Dendrocalamus membranaceus* etc.

A majority of the plants are reported to have some medicinal value. Most of these uses are, however, documented in the form of ethnobotanical records. Some of these medicinal plants are *Abroma angusta*, *Achyranthes aspera*, *Bauhinia vahlii*, *Blainvillea acmella*, *Calotropis procera*, *Catharanthus pusillus*, *Desmodium laxiflorum*, *Dioscorea bulbifera*, *Elephantopus scaber*, *Euphorbia heterophylla* etc. Also, a search of ethnic knowledge revealed that a large number of wild plants are being used as edibles, like - *Abelmoschus moschatus*, *Acmella oleracea*, *Bauhinia purpurea*, *Begonia picta*, *Canavalia gladiata*, *Careya arborea*, *Dendrocalamus hamiltonii*, *Dioscorea villosa*, *Eugenia roxburghii*, *Eryngium foetidum* etc. Fuel wood demand is also part and parcel of rural and tribal dominated area, some of fuel wood species are - *Adina cordifolia*, *Albizia chinensis*, *Bauhinia divaricata*, *Bombax ceiba*, *Butea monosperma*, *Dalbergia sissoo*, *Eucalyptus tereticornis*, *Lagerstroemia parviflora*, *Mallotus philippensis* etc. Some of the fibre-yielding plants are *Agave sisalana*, *Crotalaria juncea*, *Crotalaria retusa*, *Eulaliopsis binata*, *Hibiscus mutabilis*, *Lagerstroemia parviflora* and *Linum usitatissimum*.

■■■■

CONCLUSIONS

Mekal hills are one of the most biologically diverse regions of the central India. The area is also rich in natural diversity, with diverse flora and fauna as well as in ethnic diversity. Majority the resident communities use ethnic knowledge, utilizing local plants and animals for their sustenance and economic wellbeing. Unfortunately, this knowledge is getting depleted fast, due to factors like population pressure, rise in urbanization and consequent deforestation. Therefore, it is needed to document this knowledge before it fades into oblivion. This knowledge could provide information about new drugs, food and forage species. Digital database of this traditional knowledge is a must, especially, in form of peoples biodiversity register, to avoid biopiracy, patent infringements etc. Apart from documentation, it is need of the hour that most of the plant species should be studied in relation to their natural population status. This is particularly true for this region, as more than 80% species of the area are not assessed in IUCN red list. Hence synecological studies, regeneration rates and harvest should be monitored spatially and temporally. Population monitoring, mapping of these species by means of remote sensing and GIS is also important. In central India, much of the forest is dominated by deciduous species like *Shorea robusta* (Sal). This species contributes in plenty to the leaf litter in dry season which is burnt uncontrollably during the dry season. The resulting forest fire threatens plant and animals life of the area. There are many issues to be dealt with while managing our landscapes and forests. The work carried out in this phase is certainly not conclusive and a regular follow up is required to bring affront the entire floristic diversity of this excellent landscape. We can hope to continue this work in times to come, so that a complete picture of the floristic wealth of the region emerges out.

■■■■

REFERENCES

Takhtajan, A. 1969: *Flowering Plants, Origin and Dispersal.* Tr. Jeffery, Edinburgh.

Singh, P. and Dash, S.S. 2014: *Plant Discoveries 2013* – New Genera, Species and New Records. Botanical Survey of India, Kolkata.

Roy, G.P., Shukla, B.K. and Datt, B. 1992: *Flora of Madhya Pradesh.* Ashish Publishing House. 1-639.

Khemikar I. 2000: *Common Indian Wild Flowers.* Bombay Natural History Society and Oxford University Press. 1-141.

Polunin, O. and Stainton, A. 1984: *Flowers of the Himalaya.* Oxford University Press. 1-580.

Krishen, P. 2014: *A field guide for tree spotters: Jungle trees of Central India.* Penguin Books India Private Limited. 1-400.

Mahapatra, A.K. and Panda, P.C. 2009: *Wild edible fruit plants of Eastern India.* Regional Plant Resource Center, Bhubaneshwar. 1-341.

■■■■

Colour Photo Pages

1. *Abelmoschus moschatus*

2. *Abroma angusta*

3. *Acacia auriculiformis*

4. *Acalypha hispida*

5. *Acalypha wilkesiana*

6. *Achyranthes aspera*

7. *Acmella oleracea*

8. *Adenanthera microsperma*

9. *Adina cordifolia*

10. *Aegle marmelos*

11. *Aeschynomene americana*

12. *Aeschynomene indica*

13. *Afrohybanthus enneaspermus*

14. *Agave sisalana*

15. *Ageratum conyzoides*

16. *Albizia chinensis*

17. *Albizia lebbeck*

18. *Albizia odoratissima*

19. *Albizia procera*

20. *Aloe vera*

21. *Alstonia scholaris*

22. *Alternanthera brasiliana*

23. *Alternanthera ficoidea*

24. *Alysicarpus monilifer*

25. *Amaranthus hybridus*

26. *Amaranthus spinosus*

27. *Ampelocissus latifolia*

28. *Andrographis paniculata*

29. *Anisomeles indica*

30. *Annona squamosa*

31. *Anogeissus latifolia*

32. *Argemone mexicana*

33. *Argyreia nervosa*

34. *Arisaema tortuosum*

35. *Arthraxon hispidus*

36. *Artocarpus heterophyllus*

37. *Asparagus racemosa*

38. *Averrhoa carambola*

39. *Azadirachta indica*

40. *Baccharoides anthelmintica*

41. *Bambusa vulgaris*

42. *Barleria prattensis*

43. *Bauhinia divaricata*

44. *Bauhinia purpurea*

45. *Bauhinia racemosa*

46. *Bauhinia vahlii*

47. *Begonia picta*

48. *Bidens pilosa*

49. *Biophytum sensitivum*

50. *Blainvillea acmella*

51. *Boerhavia diffusa*

52. *Bombax ceiba*

53. *Bonnaya ciliata*

54. *Bougainvillea spectabilis*

55. *Brassica rapa*

56. *Bridelia retusa*

57. *Buchanania cochinchinensis*

58. *Butea monosperma*

59. *Caesalpinia bonduc*

60. *Caesalpinia decapetala*

61. *Caesalpinia pulcherrima*

62. *Caesulia axillaris*

63. *Cajanus cajan*

64. *Cajanus scarabaeoides*

65. *Caladium bicolor*

66. *Calliandra haematocephala*

67. *Callistemon citrinus*

68. *Calotropis procera*

69. *Canavalia gladiata*

70. *Canna indica*

71. *Cardamine hirsuta*

72. *Careya arborea*

73. *Carissa spinarum*

74. *Caryota urens*

75. *Cascabela thevetia*

76. *Catharanthus pusillus*

77. *Catunaregam spinosa*

78. *Celosia argentea*

79. *Celosia spicata*

80. *Cenchrus biflorus*

81. *Cestrum nocturnum*

82. *Chamaecrista mimosoides*

83. *Chionachne gigantea*

84. *Chloris radiata*

85. *Chrozophora rottleri*

86. *Chrysopogon aciculatus*

87. *Cinnamomum camphora*

88. *Cleome monophylla*

89. *Clerodendrum indicum*

90. *Clitoria ternatea*

91. *Cocos nucifera*

92. *Codiaeum variegatum*

93. *Colebrookea oppositifolia*

94. *Colocasia esculenta*

95. *Combretum indicum*

96. *Commelina diffusa*

97. *Convolvulus equitans*

98. *Convolvulus prostratus*

99. *Corchorus fascicularis*

100. *Cordia myxa*

101. *Corynandra viscosa*

102. *Cosmos bipinnatus*

103. *Cosmos caudatus*

104. *Cosmos sulphureus*

105. *Crassocephalum crepidioides*

106. *Crateva magna*

107. *Crinum asiaticum*

108. *Crotalaria calycina*

109. *Crotalaria juncea*

110. *Crotalaria medicaginea*

111. *Crotolaria pallida*

112. *Crotalaria retusa*

113. *Crotalaria spectabilis*

114. *Cucumis callosus*

115. *Cullen corylifolium*

116. *Cuphea carthagenensis*

117. *Cuphea hyssopifolia*

118. *Curculigo orchioides*

119. *Curcuma caesia*

120. *Cyanotis vaga*

121. *Cynodon dactylon*

122. *Cynoglossum wallichii*

123. *Cyperus difformis*

124. *Cyperus esculentus*

125. *Cyperus iria*

126. *Cyperus rotundus*

127. *Dactyloctenium aegyptium*

128. *Dalbergia sissoo*

129. *Datura metel*

130. *Datura stramonium*

131. *Delonix regia*

132. *Dendrocalamus hamiltonii*

133. *Dendrocalamus membranaceus*

134. *Desmodium laxiflorum*

135. *Dianthus woroschilovii*

136. *Dichanthium annulatum*

137. *Dichrostachys cinerea*

138. *Digitaria ciliaris*

139. *Digitaria eriantha*

140. *Dinebra retroflexa*

141. *Dioscorea bulbifera*

142. *Dioscorea villosa*

143. *Diospyros melanoxylon*

144. *Diplocyclos palmatus*

145. *Dombeya wallichii*

146. *Duranta erecta*

147. *Echinochloa colona*

148. *Eclipta prostrata*

149. *Eichhornia crassipes*

150. *Elaeocarpus ganitrus*

151. *Elaeodendron glaucum*

152. *Elephantopus scaber*

153. *Eleusine indica*

154. *Emilia sonchifolia*

155. *Epiphyllum oxypetalum*

156. *Eragrostis amabilis*

157. *Eragrostis minor*

158. *Erigeron bonariensis*

159. *Eryngium foetidum*

160. *Eucalyptus tereticornis*

161. *Eugenia roxburghii*

162. *Eulaliopsis binata*

163. *Euphorbia cotinifolia*

164. *Euphorbia heterophylla*

165. *Euphorbia hirta*

166. *Euphorbia milii*

167. *Euphorbia pulcherrima*

168. *Euphorbia thymifolia*

169. *Evolvulus alsinoides*

170. *Evolvulus nummularius*

171. *Ficus benghalensis*

172. *Ficus elastica*

173. *Ficus hispida*

174. *Ficus mollis*

175. *Ficus palmata*

176. *Ficus racemosa*

177. *Ficus religiosa*

178. *Ficus semicordata*

179. *Fimbristylis littoralis*

180. *Fimbristylis quinquangularis*

181. *Flemingia macrophylla*

182. *Gaillardia grandiflora*

183. *Galphimia speciosa*

184. *Gardenia jasminoides*

185. *Gardenia resinifera*

186. *Garuga pinnata*

187. *Glinus oppositifolia*

188. *Gloriosa superba*

189. *Gmelina arborea*

190. *Gomphrena celosioides*

191. *Gossypium arboreum*

192. *Grangea maderaspatana*

193. *Grevillea robusta*

194. *Grewia hirsuta*

195. *Grewia oppositifolia*

196. *Guizotia abyssinica*

197. *Habenaria marginata*

198. *Habenaria roxburghii*

199. *Hallenia speciosa*

200. *Hedychium coronarium*

201. *Helianthus strumosus*

202. *Helicteres isora*

203. *Heliotropium supinum*

204. *Hemerocallis fulva*

205. *Hibiscus mutabilis*

206. *Hibiscus rosa-sinensis*

207. *Hibiscus sabdariffa*

208. *Hibiscus syriacus*

209. *Hydrangea macrophylla*

210. *Hygrophila schulli*

211. *Impatiens balsamina*

212. *Indigofera astragalina*

213. *Indigofera hirsuta*

214. *Indigofera linifolia*

215. *Ipomoea aquatica*

216. *Ipomoea carnea*

217. *Ipomoea eriocarpa*

218. *Ipomoea hederifolia*

219. *Ipomoea nil*

220. *Ipomoea pes-tigridis*

221. *Ixora chinensis*

222. *Jacaranda mimosifolia*

223. *Jasminum sambac*

224. *Jatropha curcas*

225. *Jatropha gossypiifolia*

226. *Justicia adhatoda*

227. *Justicia procumbens*

228. *Kalanchoe blossfeldiana*

229. *Kalanchoe pinnata*

230. *Kyllingia brevifolia*

231. *Lablab purpureus*

232. *Lagascea mollis*

233. *Lagerstroemia indica*

234. *Lagerstroemia parviflora*

235. *Lagerstroemia speciosa*

236. *Lantana camara*

237. *Leea asiatica*

238. *Leea rubra*

239. *Leonotis nepetifolia*

240. *Leucaena leucocephala*

241. *Leucas aspera*

242. *Leucas cephalotes*

243. *Limonia acidissima*

244. *Lindernia dubia*

245. *Linum usitatissimum*

246. *Ludwigia perennis*

247. *Madhuca longifolia*

248. *Magnolia champaca*

249. *Mallotus philippensis*

250. *Malva sylvestris*

251. *Mangifera indica*

252. *Manilkara zapota*

253. *Mansoa alliacea*

254. *Martynia annua*

255. *Mecardonia procumbens*

256. *Melochia corchorifolia*

257. *Mentha piperita*

258. *Merremia emarginata*

259. *Mesosphaerum suaveolens*

260. *Mimosa pudica*

261. *Mirabilis jalapa*

262. *Mitragyna parvifolia*

263. *Moringa oleifera*

264. *Mucuna pruriens*

265. *Murdannia nudiflora*

266. *Murraya koenigii*

267. *Murraya paniculata*

268. *Musa paradisiaca*

269. *Mussaenda frondosa*

270. *Nelumbo nucifera*

271. *Neolamarckia cadamba*

272. *Nerium indicum*

273. *Nicandra physalodes*

274. *Nyctanthes arbor-tristis*

275. *Nymphaea lotus*

276. *Nymphoides indica*

277. *Ocimum basilicum*

278. *Ocimum tenuiflorum*

279. *Oligochaeta ramosa*

280. *Opuntia stricta*

281. *Ottelia alismoides*

282. *Oxalis corniculata*

283. *Oxalis latifolia*

284. *Panicum maximum*

285. *Parasopubia delphinifolia*

286. *Parthenium hysterophorus*

287. *Paspalidium flavidum*

288. *Paspalum distichum*

289. *Paspalum scrobiculatum*

290. *Peltophorum pterocarpum*

291. *Pennisetum hohenackeri*

292. *Persicaria glabra*

293. *Petalidium oblongifolium*

294. *Phoenix sylvestris*

295. *Phyllanthus emblica*

296. *Phyllanthus fraternus*

297. *Phyllanthus reticulatus*

298. *Phyllodium pulchellum*

299. *Physalis lagascae*

300. *Piper longum*

301. *Pithecellobium dulce*

302. *Plectranthus barbatus*

303. *Plectranthus mollis*

304. *Plectranthus scutellarioides*

305. *Plumbago zeylanica*

306. *Plumeria alba*

307. *Plumeria pudica*

308. *Polianthes tuberosa*

309. *Polyalthia longifolia*

310. *Polygala arvensis*

311. *Polygonum plebejum*

312. *Pongamia pinnata*

313. *Pontederia cordata*

314. *Psidium guajava*

315. *Putranjiva roxburghii*

316. *Pycreus polystachyos*

317. *Pyrus communis*

318. *Radermachera xylocarpa*

319. *Rauvolfia serpentina*

320. *Rhaphidophora decursiva*

321. *Ricinus communis*

322. *Rosa indica*

323. *Rotheca serrata*

324. *Rubia cordifolia*

325. *Ruellia tuberosa*

326. *Rumex dentatus*

327. *Saccharum spontaneum*

328. *Salvia splendens*

329. *Sapindus mukorossi*

330. *Saraca indica*

331. *Scoparia dulcis*

332. *Semecarpus anacardium*

333. *Senegalia pennata*

334. *Senna alata*

335. *Senna auriculata*

336. *Senna occidentalis*

337. *Senna siamea*

338. *Senna tora*

339. *Sesamum indicum*

340. *Sesbania sesban*

341. *Setaria helvola*

342. *Shorea robusta*

343. *Sida acuta*

344. *Sida cordifolia*

345. *Sida rhombifolia*

346. *Smithia conferta*

347. *Solanum nigrum*

348. *Solanum torvum*

349. *Solanum viarum*

350. *Solanum virginianum*

351. *Sonchus oleraceus*

352. *Spathodea campanulata*

353. *Spermacoce articularis*

354. *Spermacoce hispida*

355. *Spermacoce ocymoides*

356. *Sphaeranthus indicus*

357. *Spilanthes acmella*

358. *Spondias pinnata*

359. *Stachytarpheta jamaicensis*

360. *Stephania japonica*

361. *Stereospermum colais*

362. *Synedrella nodiflora*

363. *Syzygium cumini*

364. *Tabernaemontana divaricata*

365. *Tagetes erecta*

366. *Tagetes patula*

367. *Tamarindus indica*

368. *Tecoma stans*

369. *Tectona grandis*

370. *Terminalia alata*

371. *Terminalia arjuna*

372. *Terminalia bellirica*

373. *Terminalia chebula*

374. *Tetrastigma leucostaphylum*

375. *Themeda quadrivalvis*

376. *Thespesia lampas*

377. *Thespesia populnea*

378. *Thunbergia erecta*

379. *Thunbergia fragrans*

380. *Thysanolaena latifolia*

381. *Tithonia diversifolia*

382. *Toona ciliata*

383. *Tradescantia pallida*

384. *Trema orientalis*

385. *Trichodesma zeylanicum*

386. *Trichosanthes tricuspidata*

387. *Tridax procumbens*

388. *Triumfetta rhomboidea*

389. *Turnera ulmifolia*

390. *Typha angustifolia*

391. *Uraria lagopodoides*

392. *Uraria picta*

393. *Urena lobata*

394. *Urochloa panicoides*

395. *Utricularia aurea*

396. *Vachellia farnesiana*

397. *Vachellia nilotica*

398. *Verbascum coromandelianum*

399. *Vernonia cinerea*

400. *Vinca rosea*

401. *Vitex negundo*

402. *Volkameria inermis*

403. *Wendlandia heynei*

404. *Woodfordia fruticosa*

405. *Xanthium strumarium*

406. *Zingiber zerumbet*

407. *Ziziphus mauritiana*

408. *Ziziphus oenopolia*

409. *Ziziphus rugosa*

410. *Zornia gibbosa*

Related Books

ISBN: 9789387307100

ISBN: 9789387893894

ISBN: 9789389061833

ISBN: 9789386102836

ISBN: 9789383692552

ISBN: 9789386102768

ISBN: 9789386652263

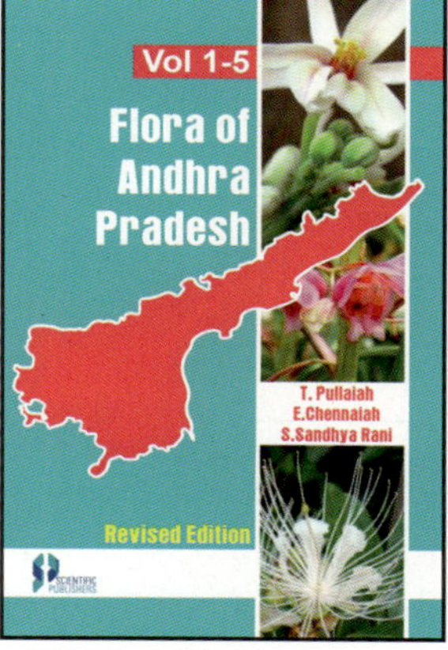

ISBN: 9789386347954

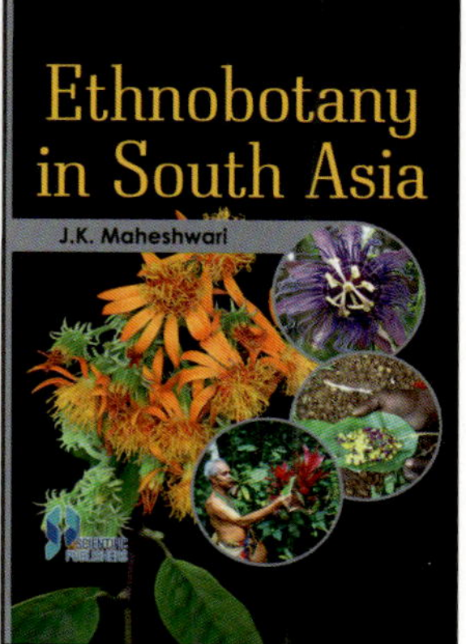

ISBN: 9788172331436

Related Books

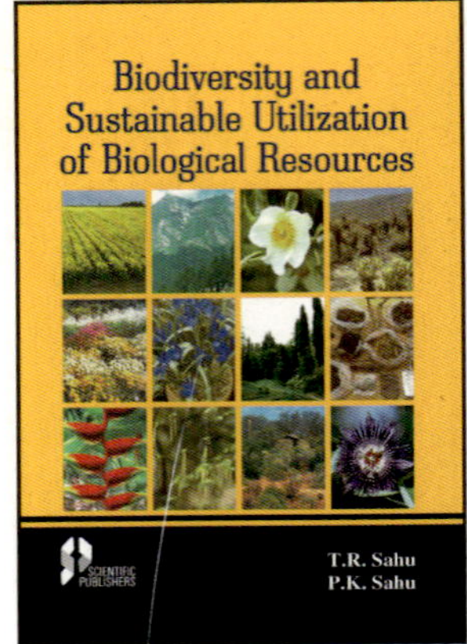

ISBN: 9788172333621

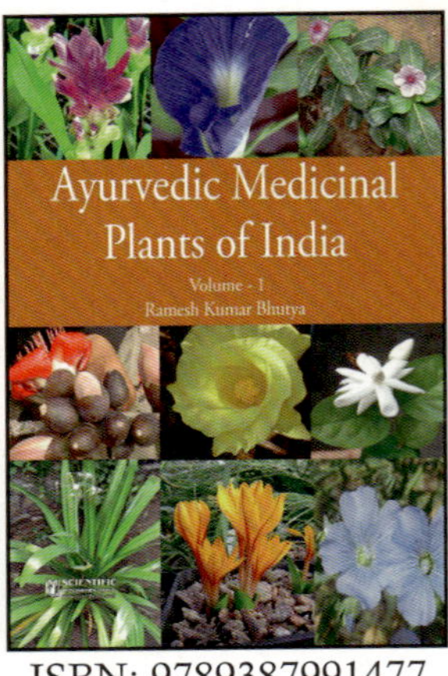

ISBN: 9789387991477

ISBN: 9788172338176

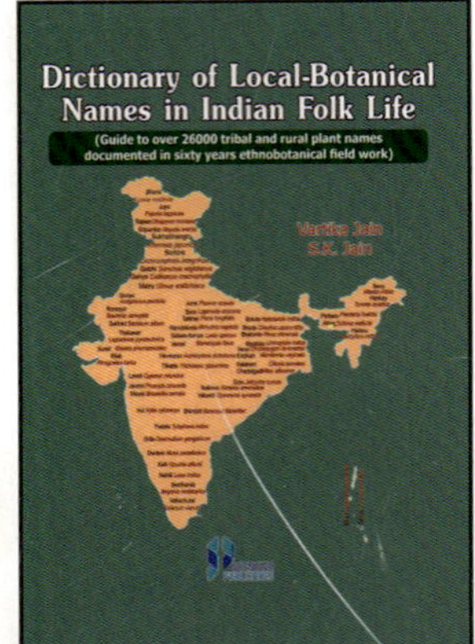

ISBN: 9789383692514

ISBN: 9788172337766

ISBN: 9788172338565

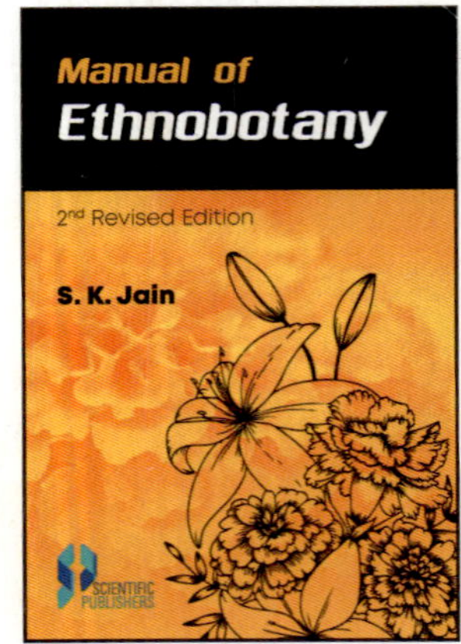

ISBN: 9788172336530

ISBN: 9788172339753

ISBN: 9789386652195